Cartilage Tissue Engineering and Regeneration Techniques

Edited by Dimitrios D. Nikolopoulos, George K. Safos and Kalpaxis Dimitrios

Published in London, United Kingdom

IntechOpen

Supporting open minds since 2005

Cartilage Tissue Engineering and Regeneration Techniques
http://dx.doi.org/10.5772/intechopen.77717
Edited by Dimitrios D. Nikolopoulos, George K. Safos and Kalpaxis Dimitrios

Contributors
Lourenço Heitzmann, Samina Haq, Iqraa Haq, Atheer AliAlsayahi, Abir Alamro, Amani Alghamedi, Lizeth Fuentes-Mera, Elisabeth Engel, Alberto Camacho, Vanessa Pérez-Silos, Jorge Lara-Arias, Ivan Marino-Martínez, Victor Peña-Martinez, Keolebogile Shirley Caroline Motaung, Makwese Johannes Maepa, Mapula Razwinani, Franca NNeka Alaribe, Jianying Zhang, He Huang, Hongyao Xu, Carlos Semino, Nausika Betriu, José Antonio A. Vega, Juan L. Cobo, Manuela Cabrera-Freitag, Teresa Cobo, Juan D. Muriel, Juan Cobo, Marcelo Sa, Francisco Alipio Sousa Segundo, Roberto Lopes De Souza

First published in London, United Kingdom, 2019 by IntechOpen
IntechOpen is the global imprint of INTECHOPEN LIMITED, registered in England and Wales, registration number: 11086078, The Shard, 25th floor, 32 London Bridge Street
London, SE19SG - United Kingdom
Printed in Croatia

British Library Cataloguing-in-Publication Data
A catalogue record for this book is available from the British Library

Additional hard and PDF copies can be obtained from orders@intechopen.com

Cartilage Tissue Engineering and Regeneration Techniques
Edited by Dimitrios D. Nikolopoulos, George K. Safos and Kalpaxis Dimitrios
p. cm.
Print ISBN 978-1-78923-849-5
Online ISBN 978-1-78923-850-1
eBook (PDF) ISBN 978-1-83880-781-8

For EU product safety concerns:
IN TECH d.o.o., Prolaz Marije Krucifikse Kozulić 3, 51000 Rijeka, Croatia, info@intechopen.com or visit our website at intechopen.com.

Meet the editors

Dr. Dimitrios D. Nikolopoulos is a graduate of Athens University Medical School. He specialized in orthopedic surgery and traumatology in one of the best clinics in Greece, the Orthopaedic Clinic of the General Hospital "ASKLEPIEION" Voulas, where he focused on sports injuries and foot pathology. He proceeded to train alongside Dr. Alexis Nogier at the Clinic Maussins-Nollet in Paris with arthroscopic restoration of hip pathology and has continued his training in the arthroscopic rehabilitation of sports injuries of the knee, shoulder, and ankle, as well as the treatment and surgical correction of foot disorders. He is a Member of the International Cartilage Regeneration Society. Dr. Nikolopoulos and Dr. G. Safos have developed new techniques for ankle cartilage defects presented and published internationally.

Dr. George K. Safos is a graduate of Athens University Medical School and completed his residency in orthopedic surgery and traumatology at the General Hospital of Asklepieion Voulas. He then completed two fellowships in the United States, namely at the National Institutes of Health in Bethesda, Maryland, and at the University of Miami-Orthopedic Department. A postgraduate degree in Tissue Banking at the National University of Singapore with a United Nations scholarship soon followed. Additionally, he completed a prestigious foot and ankle course at the Academic Centre of Amsterdam University and many more internationally renowned postgraduate courses. Past work experience includes doctor of the Olympiacos basketball team for five years and orthopedic surgeon in state, military, and private hospitals alike. Dr. Safos and Dr. D. Nikolopoulos have developed new techniques for ankle cartilage defects and have presented and published internationally.

Kalpaxis Dimitrios (MD) graduated from the Athens Technological Institute (Department of Physiotherapy) in 2007 and has an MSc in Metabolic Bone Disease (Faculty of Medicine, Athens National and Kapodistrian University). His internship was completed at St. Panteleimon General Hospital in Nikea (Piraeus). During his military service, he worked as a physiotherapist for the Athens Navy Hospital (2008). Kalpaxis also worked as a physiotherapist for the Olympiacos water polo team, the Hippokration General Hospital of Athens, and the "GALILEE" Palliative Care Unit. He has followed some of the most important seminars and is a member of the Panhellenic Association of Physiotherapists. From 2011, he has been a member of the Physiotherapy Team of the Central Clinic of Athens.

Contents

Preface

Joint injuries and the resultant articular cartilage defects are frequent occurrences that an orthopedic surgeon/arthroscopist faces every day. Progressive wear and tear of articular cartilage can lead to progressive cartilage tissue loss, further exposing the bony ends, leaving them without protection. This final osteochondral defect deteriorates latterly into the most common arthritis, a degenerative joint disease. Unlike other self-repairing tissues, such as bone, cartilage has a low regenerative capacity. As a consequence, once injured, cartilage is much more difficult to self-heal. The field of articular cartilage tissue engineering, which aims to repair, regenerate, and/or improve injured or diseased articular cartilage functionality, has evoked intense interest and held great potential for improving articular cartilage therapy over the last decade. This book provides an overall description of classic and current articular cartilage repair and regeneration therapies and techniques, conventional or operative. Also, it summarizes recent progress in cartilage tissue engineering, including stem cells, growth factors, bioactive molecules, and biomaterial scaffolds used for cartilage regeneration. Finally, the technical and regulatory challenges of articular cartilage tissue engineering and possible future directions are discussed.

Dr. Dimitrios D. Nikolopoulos (MD, PhD)
Orthopaedic Surgeon, Arthroscopist,
Central Clinic of Athens, Greece
Member of Hellenic Association of Orthopaedic Surgery and Traumatology,
Hellenic Arthroscopic Association, ESSKA, ESSKA-AFAS,
AO Trauma, OTA, ICRS,
Athens, Greece

Dr. George K. Safos (MD, PhD)
Orthopaedic Surgeon, Arthroscopist,
Central Clinic of Athens, Greece
Member of Hellenic Association of Orthopaedic Surgery and Traumatology,
ESSKA,
Athens, Greece

Mr. Kalpaxis Dimitrios (MD)
Physiotherapist,
Central Clinic of Athens,
Greece

Chapter 1

Cartilage Tissue Engineering and Regeneration

Francisco Sousa Segundo, Marcelo Jorge Cavalcanti de Sá and Roberto Lopes de Souza

Abstract

Cartilaginous tissue is mainly present in the joints, and it consists predominantly of type II collagen and glycoproteins, which promote functions of supporting biomechanical forces generated during the ambulation. The cartilage has a very limited regenerating capacity, causing traumas or degenerative diseases in this region difficult to solve. The current treatments for regeneration of the articular cartilages may be conservative or surgical, but they are not very successful, since the damaged tissue is replaced by fibrous tissue or fibrocartilage, with predominantly type I collagen, which present inferior functions. Cellular therapies, biomaterials, and tissue engineering to assist the healing process have been showing great potential. For example, the in vitro chondrogenesis of mesenchymal stem cells (MSCs) is a technique that stimulates undifferentiated cells to transform into chondrocytes, creating a dense mass of aggregated MSCs and an environment with strong cell-cell interactions.

Keywords: biomaterials, cell therapy, joint, tissue repair, regenerative medicine

1. Introduction

Joint diseases such as osteoarthritis can cause important lesions in the articular cartilage; in humans, this pathology can affect a significant proportion of patients over 60 years of age, causing a great negative impact on their quality of life. This increased the search for an effective treatment the objective of several researches, often seeking the cooperation of several areas of knowledge; however the difficulty in repairing articular defects in an effective way is becoming a real challenge for medicine [1].

The cartilaginous tissue present in the joints is a highly organized and specialized tissue, presenting several fundamental mechanical proprieties for the maintenance of articular function [1].

The lesions caused in this tissue from trauma or degenerative diseases cause a gradual damage to the tissue, leading to joint pain and consequent impairment in its function, which are difficult to handle clinically [1].

Thus, cases of severe joint disease are usually treated surgically, either through osteotomies and the application of autologous subchondral grafts, reducing the progression of the cartilage lesion and promoting return to joint function, or, in more severe cases, complete articular replacement through prosthesis implantation [2].

Due to the fact that these techniques may present unsatisfactory clinical results, alternative treatments for repair of this type of injury are constant aims of research, mainly in the area of regenerative medicine and tissue engineering, which are commonly working together for the development of different therapies [2].

With the development of regenerative techniques, such as the use of biomaterials and implantation of autologous cells or tissue, promising results emerged; it was also noticed that the association of several technical modalities was indispensable for obtaining satisfactory clinical recoveries [3].

2. Cartilage proprieties

The articular cartilage is formed basically of hyaline cartilage, being rich in type II collagen fibers and glycoproteins; this tissue is present in the end of the long bones and sesamoids with synovial articulation, as well as in the physeal line, which divides the diaphysis and epiphysis and is responsible for part of the bone growth [4].

The healthy articular cartilage has a macroscopic aspect with shiny and whitish color and smooth surface, is always bathed by the synovial fluid which is produced by the synovial membranes, and is contained in the articular capsule. Its main function is the cartilage nutrition and articular lubrication, and its translucent and viscose consistency contributes to the sliding of the articular segments and reduction of the articular friction during the movement [5].

Histologically the articular cartilage is defined as a highly specialized tissue with cells called chondrocytes, which produce collagen fibers, distributed in rows, in the periphery of the tissue; these cells present in elliptical form with the largest axis parallel to the tissue surface, in the center of the cartilage, and have a rounded shape and can group together. Another component present in the cartilaginous tissue is the amorphous substance which is composed of macromolecules of glycosaminoglycans [6].

This tissue has little or no vascularization, and its nutritional supply comes almost exclusively from the synovial fluid. This makes it difficult for cell migration and proliferation in injured sites; due to this, treatments through microfractures, where a surgical perforation is performed in the cartilaginous lesion aiming to cause hemorrhage and clot formation from the subchondral bone, are widely employed. Despite being one of the most used techniques, clinical results are often unsatisfactory [7].

The cartilaginous tissue has almost no innervation, which favors its function of capturing and distributing mechanical forces applied to it during movement, without any painful reaction. Full-thickness lesions in the cartilage cause the subchondral bone tissue lying just below the cartilage to be exposed, and the friction generated during movement causes contact with the subchondral bone to cause a debilitating pain for joint function [8].

The impairment of the normal function of a joint can negatively influence several nearby structures, and the lack of adequate use of a limb with joint problems leads to atrophy of muscle groups, which in addition to having clinical impacts on patients' quality of life can still make it impossible for some patients to work [8].

2.1 Cartilage lesions

Initially it was believed that osteoarthritis (OA) occurred due to wear and tear of the articular surface; however, it is now understood that the development process of such pathology has much more complex mechanisms [9].

The changes observed in the joints affected with OA are usually directly related to the severity of the same, being observed formation of osteophytes, inflammation of articular and periarticular components, bone deformity, and degeneration of the cartilaginous tissue, as shown in **Figure 1** [11].

Several causes can lead to impairment of the normal function of articular structures, including trauma and degenerative diseases, culminating mainly in mechanical instabilities, which promote cartilage surface abrasion and progressive degeneration [12]. This is due to the release of cellular communicators such as cytokines; interleukin types 1, 4, 9, and 13; and tumor necrosis factor alpha (TNF-a), as well as release of enzymes such as disintegrin, metalloproteinase thrombospondin-like motifs, and collagenases; all this activity is carried out by chondrocytes, osteoblast, and synoviocytes, as shown in **Figure 2** [14]. It is also believed that the innate immune system may participate in the progression of OA, mainly by the activation of the complement and alternative pathways [15].

After stimulation, the release of enzymes in the joint leads to degradation of the cartilaginous matrix, causing chondrocyte hypertrophy, which loses the ability to produce a new collagen matrix [16]. The proliferation of the subchondral bone exceeds the intersection between the bone and cartilage, causing the formation of osteophytes, subchondral cysts, and subchondral sclerosis; all of this process aims to compensate for a mechanical instability in the joint and redistribute the forces acting on it [17].

However, the continuous exposure of the tissue to mechanical stimuli leads to the release of vascular endothelial growth factors by chondrocytes, which promote intense neoangiogenesis and invasion of the joint tissues. Patients affected by angular deviations or instabilities that increase joint exposure to poor distribution of mechanic forces have a much more aggressive progression of OA, with subchondral bone damage associated with severe articular pain [18].

In these cases pain may be associated with the remodeling of the subchondral bone, which is widely innervated, and the inflammatory process of the joint structures may contribute to the aggravation of pain; in more chronic cases, the joint membrane may become fibrosed, which compromises the performance of its

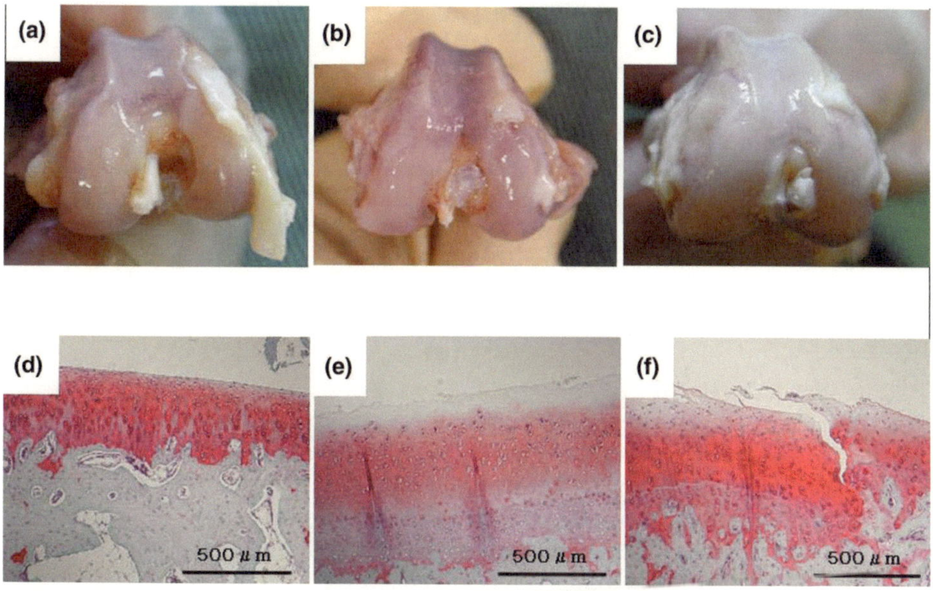

Figure 1.
Macroscopic findings of rabbit articular cartilage at 1 week (a), 4 weeks (b), and 8 weeks (c) after collagenase injection. Photomicrographs of rabbit articular cartilage at 1 week (d), 4 weeks (e), and 8 weeks (f) after collagenase injection are also shown (Safranin O staining; magnification × 4); taken with permission [10].

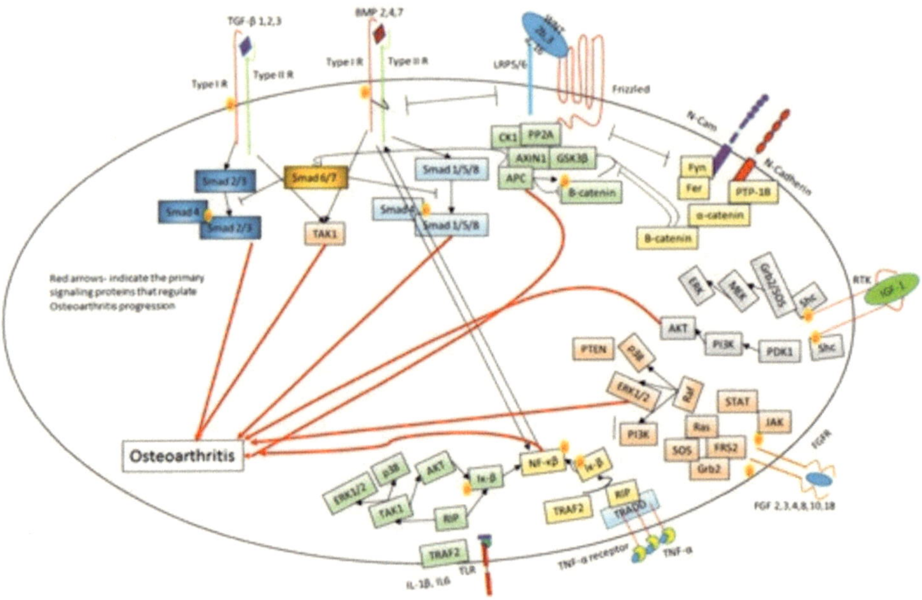

Figure 2.
Signaling cascades involved in osteoarthritis. Red arrows indicate the primary signaling protein that regulates OA progression. The black arrows signify the activation of the proteins. The bars indicate inhibition of the proteins; taken with permission [13].

normal functions, and analgesic treatment is of great importance and can be difficult if central or peripheral sensitization occurs [19].

The development of therapeutic modalities for the management of OA is a constant interest of researchers; commonly experimental animal models are used to investigate different treatment options, and among the most used animals are guinea pigs, rabbits, rats, dogs, sheep, and horses, studying OA in different joints, such as the temporomandibular, metacarpophalangeal, and the most frequently used knee joint [20].

The development of experimental models for OA is of fundamental importance for the understanding of the pathophysiology of the pathology, for example, through experimental models in rats as shown in **Figure 3**; it was possible to observe a much more complex relation between the histological morphology of the cartilaginous tissue and the pain phenotypes. It can contribute significantly in understanding the mechanisms of cellular development and interaction in diseased joints as well as in the targeting of patients' analgesic therapy [21].

2.2 Tissue engineering

Complications related to the surgical treatment of microfractures have caused the search for alternatives or associations to grow rapidly; one of the areas with promising results is tissue engineering, which cultivates cells or tissues in vitro, intended for implantation in injured sites [22]. Usually chondrocyte cells are collected from the patient, in joints that are not submitted to biomechanical loads; then the cells are cultured and multiplied in the laboratory, reaching amounts of 12 to 48 million cells, and then implanted in affected joints [23].

The main benefits of this technique are the low rate of rejection or complications due to foreign materials to the patient, since the base cells are collected by biopsy and are autologous to the patient and the noninvasive nature and since the

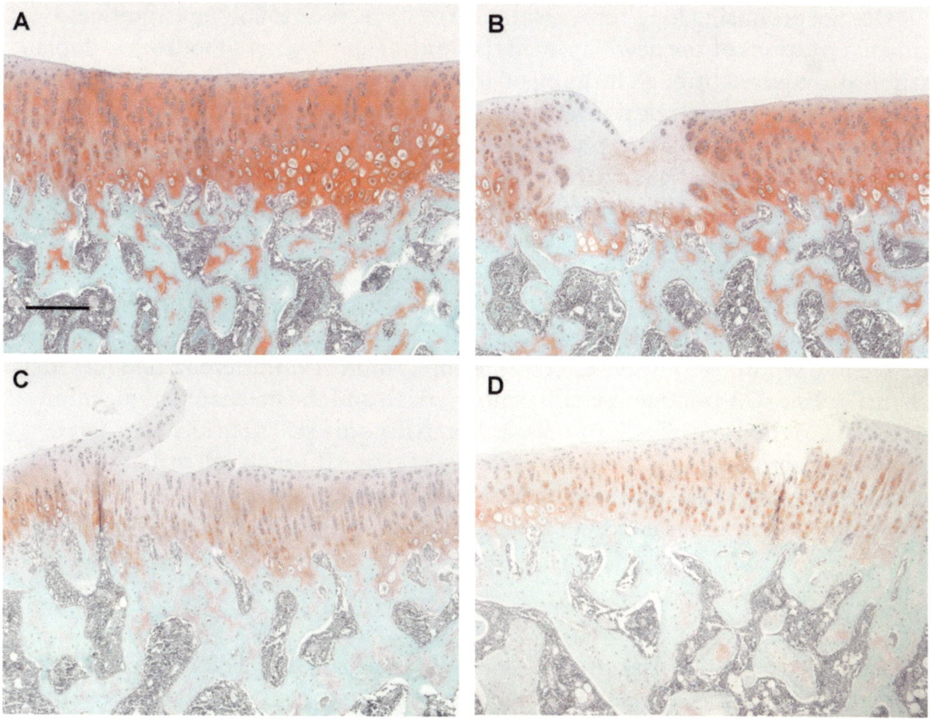

Figure 3.
Articular cartilage pathology 20 and 42 days after OA induction. Histological images of the tibial plateau (A–D). Joints were sectioned in a frontal plane and stained with Safranin O/fast green staining and corresponding consecutive sections stained with H&E. (A) Saline-treated control showing smooth cartilage and normal joint margin and chondrocyte morphology. (B–C) 1 mg and 0.1 mg glycolysis inhibitor monosodium iodoacetate-injected rats at day 20 and (D) 0.1 mg glycolysis inhibitor monosodium iodoacetate-injected rat at day 42 show degeneration of the cartilage. Proteoglycan loss (B–D) and chondrocyte cloning (B) are also present in the arthritic cartilage. Scale bars = 200 μm. Images are of knees with median cartilage surface integrity scores from each group; taken with permission [21].

collection and implantation are usually performed through arthroscopy, which promotes greater comfort and better recovery for the tissue [24].

The main negative points observed in this technique are the need for two interventions, a relatively long recovery time, between 6 and 12 months, which are necessary for the tissue neoformation and its maturation, and the intrinsic complexity of performing this procedure [25].

Clinical studies and follow-up of long-term cases have confirmed excellent clinical and functional results of patients submitted to autologous chondrocyte grafting in articular defects of up to 4 cm^2 [26]; positive results were also observed when associated with grafting technique with corrective osteotomy [27]; when compared with the microfracture techniques, the autologous chondrocyte grafting obtained superior results in joint defects of 3 cm^2 or greater, entertained, and when these two techniques were used in minor defects, no difference was observed between the treatments [28].

The autogenous chondrocyte graft appears to have limitations, especially in relation to patients with very large articular defects >15 cm^2 and with severe tissue lesions; in these patients the graft technique is associated with a low survival rate of the implanted tissue, and despite this, no clinical differences were observed with patients whose grafts were successful [26].

In humans the autologous chondrocyte implant seems to have a more favorable result when used in younger patients and with a joint disease in a period shorter than 12 months, with a rate of return of high impact sports of up to 96% of patients treated [29].

Despite promising long-term results, up to 10 years of follow-up, additional characterizations of the newly formed tissue after autologous chondrocyte implantation are required through imaging or arthroscopy [30].

An alternative to surgical implantation of tissue manufactured in vitro is the intra-articular application of pluripotent cells, such as allogeneic or autologous undifferentiated mesenchymal cells; these are derived from bone marrow, adipose tissue, or the umbilical cord [31]. This technique has been used mainly for the purpose of reducing joint pain and reducing the progression of tissue degeneration; between 8 and 9 million autologous cells are administered per patient or 40 million allogeneic cells per patient; administrations are performed in the affected joint and promote stimulation of chondrogenesis [32].

Despite reports of positive effects, the application of undifferentiated mesenchymal cells to OA treatment is still controversial, and the mechanisms of action of the use of this type of cells in a diseased cartilage are still not fully elucidated, suggesting both direct effect on the recovery of tissue through cell differentiation and indirect effect through the release of inducing factors and tissue growth, the two actions being associated with beneficial effects on the joint [33].

2.3 Biomaterials

The advances in material engineering in association with medicine have enabled new technologies to be developed to help treat various pathologies that were real challenges. Among the technologies developed, the manufacture of materials with various biological properties that could be implanted in different tissues was an event that brought excellent results and opened up a wide spectrum of possibilities [34].

Biomaterials must possess certain characteristics to optimize their benefits, for their use in the repair of cartilaginous tissue; besides being biocompatible and providing cellular adhesion and proliferation, they must be bioactive, biomimetic, biodegradable, and bioresponsive. These characteristics added to an adequate three-dimensional arrangement favor the environmental stimulus for the production of desirable cells, such as chondrocytes [35].

Among the materials that can be used in the manufacture of biomaterials are polymers which may be either natural or synthetic. Natural polymers have a better interaction with the implanted site, providing a more natural environment for cellular development, supporting and guiding their differentiation between several stages; however, one of its negative points is the low mechanical capacity when compared with other biomaterials [36].

Synthetic polymers can have their mechanical characteristics of controlled strength, stiffness, and degradation rate, making them quite versatile; since their biological characteristics are not desirable due to their hydrophobic properties, it is often necessary to add other materials that increase your cellular interaction [37].

A collagen-based implant, developed with three layers, made with the combination of equine collagen, magnesium, and hydroxyapatite, showed a good result in patients with large joint defects; however, the number of cases was small, suggesting that more studies are required [38]. Polyglactin-based implants associated with platelet-rich plasma and hyaluronic acid have shown promising results in the treatment of joint injuries; however, larger studies are needed [39].

In addition to the regenerative properties of biomaterials, these can still be used as controlled drug delivery systems, among which the most studied are microparticle implants manufactured from poly(lactic acid), poly(lactic-co-glycolic acid), and polycaprolactone—these synthetic polymers are made up of particles measuring above 1 μm [40].

Studies using poly(lactic acid)-based microspheres as a controlled drug delivery system were able to obtain a drug release rate of 20 to 62% in 3 months when applied in vivo; the biomaterial has shown bioactivity 2 months after intra-articular application in rats, showing potential for pain control in patients with OA [41].

The same biodegradable synthetic polymers can be used to manufacture nanoparticles, which have a size smaller than 1 um, which can also carry drugs with analgesic, anti-inflammatory, or other biomaterials with regenerative properties [42].

When compared to the microparticles, the nanoparticles have a shorter action time, being eliminated in a matter of days, as observed in the study applying microparticles and nanoparticles based on chitosan, in rats with OA [43].

Another class of biomaterials is ceramics; these are widely used in the repair of bone defects, due to their excellent properties of osteoconductivity and osteoinductivity; although they have lower mechanical characteristics than other materials, their structure can be manipulated and associated with other products [44]. By associating ceramics with type 2 bone morphogenetic protein, stimulation was observed for subchondral bone growth, as well as for the cartilaginous tissue itself [45].

2.4 Conservative therapies

The drug treatment for patients with OA has as main target the control of joint pain, which is the main reason for the patients to seek medical attention [46]. Often clinical pain arises before radiographic changes, and its etiology is not fully understood; however, the inflammatory signs produced by chondrocytes seem to play an important role for joint pain [47].

The main medications used to control joint pain are nonsteroidal anti-inflammatory drugs, opioids, or the combination of both; supplementation of vitamins, hormones, and chondroprotective medications is also widely employed [48]. Meloxicam and tramadol are drugs commonly used to control pain; however, studies have shown that long-term use of these drugs can bring several side effects, including headache, nausea, diarrhea, and urinary tract infection [49, 50].

Intra-articular administration of viscous substances can also be performed; basically all products are based on hyaluronic acid, which is produced by fibroblasts and has the function of lubricating the joint [51]. Its application showed superior results to the use of placebo in the return to joint function and reduction of pain in patients with OA [52].

Rehabilitation is a fundamental element in the recovery of joint function; this modality can be used alone or in association with other therapies. The main objective is that the cartilaginous tissue adapts through exercises that promote a regular mechanical stimulus in the joint; this process is quite slow, since this tissue can take up to 2 years to reach 75% of adaptation, contrasting muscle tissue which reaches a total adaptation at 35 weeks [1].

Even with the development of several techniques and technologies for recovering patients with joint disease, physical exercise is always present in therapeutic protocols, showing that the continuous passive movement or range of motion exercises practiced from the first day after surgical interventions is fundamental for obtaining better results [53].

Studies show that the early use of controlled exercises in OA patients promotes results superior to immobilization, contributing to decrease edema, early return to physical activity, restore range of motion, articular stability, and improve patient satisfaction with the therapeutic outcome [54].

2.5 Surgical techniques

Surgical interventions may be options for the treatment of OA, in cases in which the patient has clinical signs, and the osteotomy technique and autologous osteochondral implantation (ACI) may provide a return to joint function and minimize the progression of cartilaginous degeneration and may be associated to other therapeutical modalities such as implantation of biomaterials or tissues produced by engineering; in more chronic cases, the most commonly used surgical technique is the total replacement of the joint by prostheses; however, the long-term results of these interventions do not present acceptable clinical solutions [55].

The microfracture technique aims to perforate the subchondral bone and stimulate the migration of pluripotent cells from the bone marrow to the injured site in the cartilaginous tissue; although it is considered a gold standard technique, generally this procedure produces a fibrocartilaginous tissue, which has inferior biomechanical properties when compared to the cartilaginous tissue [56].

The fragility of the fibrocartilaginous tissue produced by the microfracture process causes this procedure to promote a limited recovery of the tissue, and the neoformed tissue usually degenerates in 18 to 24 months after the procedure due to its biomechanical characteristics [57], which compromises the long-term positive results. Initially, a decrease in lesion progression is observed up to 5 years, after which time treatment failure is expected [58].

Follow-up after treatment was of great importance, as was shown in studies comparing microfracture and autologous graft of chondrocytes with a 5-year follow-up; no significant clinical difference was observed between the groups, and despite that, the samples collected through biopsy revealed that all cases of failure after intervention and the obtained tissue had inferior quality to hyaline cartilage [59].

The osteochondral autograft transplantation (OATS) consists of the collection and implantation of an autogenous or allogeneic fragment of the bone and hyaline cartilage for repair of the defect in a joint affected by OA; this technique presents better results when used for the correction of minor defects; among its negative points, one has the morbidity of the donor caused by the collection of the material for implantation [60].

Approximately 60% of patients with OA will at some point require total joint replacement or other salvage procedures such as arthrodesis and total joint replacement to remove severely injured joint segments by prostheses; patients undergoing arthrodesis have joint surface completely removed, and the implant segments are stabilized with implants for fusion to occur [61].

In spite of the invasiveness of these procedures, patients generally adapt well to the intervention, with significant pain remission and limb function returned; however, these data seem to be related to short- to medium-term follow-up, and their long-term efficacy is questioned [62].

3. Conclusions

The OA represents a complex and recurring problem in medicine, causing pain and debilitation to patients, and its effective treatment is a real challenge.

Several therapies can be employed; however, their efficiency is variable, and it is suggested that the association of more than one therapeutic modality is the best way for a better recovery.

Conflict of interest

The authors declare no conflict of interests.

DOI: http://dx.doi.org/10.5772/intechopen.85623

Author details

Francisco Sousa Segundo[1], Marcelo Jorge Cavalcanti de Sá[1*] and Roberto Lopes de Souza[2]

1 Federal University of Campina Grande, Patos, Brazil

2 Federal University of Mato Grosso, Cuiaba, Brazil

*Address all correspondence to: mjcdesa@gmail.com

References

[1] Makris EA, Gomoll AH, Malizos KN, Hu JC, Athanasiou KA. Repair and tissue engineering techniques for articular cartilage. Nature Reviews Rheumatology. 2015;**11**:21-34. DOI: 10.1038/nrrheum.2014.157

[2] Moran CJ, Pascual-Garrido C, Chubinskaya S, Potter HG, Warren RF, Cole BJ, et al. Restoration of articular cartilage. The Journal of Bone and Joint Surgery, American Volume. 2014;**19, 96**:336-344. DOI: 10.2106/JBJS.L.01329

[3] Pestka JM, Bode G, Salzmann G, Sudkamp NP, Niemeyer P. Clinical outcome of autologous chondrocyte implantation for failed microfracture treatment of full-thickness cartilage defects of the knee joint. The American Journal of Sports Medicine. 2014;**40**(2):325-331. DOI: 10.1177/0363546511425651

[4] Pichler K, Musumeci G, Vielgut I, Martinelli E, Sadoghi P, Loreto C, et al. Towards a better understanding of bone bridge formation in the growth plate – An immunohistochemical approach. Connective Tissue Research. 2013;**54**:408-415. DOI: 10.3109/03008207.2013.828715

[5] Musumeci G, Castrogiovanni P, Leonardi R, Trovato FM, Szychlinska MA, Giunta AD, et al. New perspectives for articular cartilage repair treatment through tissue engineering: A contemporary review. World Journal of Orthopedics. 2014;**5**(2):80-88. DOI: 10.5312/wjo.v5.i2.80

[6] Musumeci G, Carnazza ML, Loreto C, Leonardi R, Loreto C. β-Defensin-4 (HBD-4) is expressed in chondrocytes derived from normal and osteoarthritic cartilage encapsulated in PEGDA scaffold. Acta Histochemica. 2012;**114**:805-812. DOI: 10.1016/j.acthis.2012.02.001

[7] Goyal D, Keyhani S, Lee EH, Hui JH. Evidence-based status of microfracture technique: A systematic review of level I and II studies. Arthroscopy. 2013;**29**(9):1579-1588. DOI: 10.1016/j.arthro.2013.05.027

[8] Becerra J, Andrades JA, Guerado E, Zamora-Navas P, López-Puertas JM, Reddi AH. Articular cartilage: Structure and regeneration. Tissue Engineering. Part B, Reviews. 2010;**16**:617-627. DOI: 10.1089/ten.teb.2010.0191

[9] Luyinu EL, Narayanan G, Nair LS, Laurencin CT. Animal models of osteoarthritis: Classification, update, and measurement of outcomes. Journal of Orthopaedic Surgery and Research. 2016;**11**:19. DOI: 10.1186/s13018-016-0346-5

[10] Hattori K, Ikeuchi K, Morita Y, Takakura Y. Quantitative ultrasonic assessment for detecting microscopic cartilage damage in osteoarthritis. Arthritis Research & Therapy. 2004. DOI: 7. DOI: 10.1186/ar1463

[11] Kraus VB, Blanco FJ, Englund M, Karsdal MA, Lohmander LS. Call for standardized definitions of osteoarthritis and risk stratification for clinical trials and clinical use. Osteoarthritis and Cartilage. 2015;**23**(8):1233-1241. DOI: 10.1016/j.joca.2015.03.036

[12] Man GS, Mologhianu G. Osteoarthritis pathogenesis – A complex process that involves the entire joint. Journal of Medicine and Life. 2014;**7**(1):37-41

[13] Akkiraju H, Nohe A. Role of chondrocytes in cartilage formation, progression of osteoarthritis and cartilage regeneration. Journal of Developmental Biology. 2015;**3**(4):177-192. DOI: 10.3390/jdb3040177

[14] Sulzbacher I. Osteoarthritis: Histology and pathogenesis. Wiener Medizinische Wochenschrift (1946). 2013;**163**(9-10):212-219. DOI: 10.1007/s10354-012-0168-y

[15] Orlowsky EW, Kraus VB. The role of innate immunity in osteoarthritis: When our first line of defense goes on the offensive. The Journal of Rheumatology. 2015;**12**(3):363-371. DOI: 10.3899/jrheum.140382

[16] Beckmann R, Houben A, Tohidnezhad M, Kweider N, Fragoulis A, Wruck CJ, et al. Mechanical forces induce changes in VEGF and VEGFR-1/sFIt-1 expression in human chondrocytes. International Journal of Molecular Sciences. 2014;**15**(9):15456-15474. DOI: 10.3390/ijms150915456

[17] Sharma L, Chmiel JS, Almagor O, Felson D, Guermazi A, Roemer F, et al. The role of varus and valgus alignment in the initial development of knee cartilage damage by MRI: The MOST study. Annals of the Rheumatic Diseases. 2013;**72**(2):235-240. DOI: 10.1136/annrheumdis-2011-201070

[18] Hayashi D, Englund M, Roemer FW, Niu J, Sharma L, Felson DT, et al. Knee malalignment is associated with an increased risk for incident and enlarging bone marrow lesions in the more loaded compartments: The MOST study. Osteoarthritis and Cartilage. 2012;**20**(11):227-233. DOI: 10.1016/j.joca.2012.07.020

[19] Malfait AM, Schnizer TJ. Towards mechanism-based approach to pain management in osteoarthritis. Nature Reviews Rheumatology. 2013;**9**(11):654-664. DOI: 10.1038/nrrheum.2013.138

[20] Lankau EW, Turner PV, Mullan RJ, Galland GG. Use of nonhuman primates in research in North America. Journal of the American Association for Laboratory Animal Science. 2014;**53**(3):278-282. DOI: 10.1007/978-3-319-64731-9_10

[21] Nwosu LN, Mapp PJ, Chaoman V, Walsh DA. Relationship between structural pathology and pain behavior in a model of osteoarthritis (OA). Osteoarthritis and Cartilage. 2016;**24**:1910-1917. DOI: 10.1016/j.joca.2016.06.012

[22] Trinh TQ, Harris JD, Siston RA, Flanigan DC. Improved outcomes with combined autologous chondrocyte implantation and patellofemoral osteotomy versus isolated autologous chondrocyte implantation. Arthroscopy. 2013;**29**:566-574. DOI: 10.1016/j.arthro.2012.10.008

[23] Macmull S, Parratt MT, Bentley G, Skinner JA, Carrington RW, Morris T, et al. Autologous chondrocyte implantation in the adolescent knee. The American Journal of Sports Medicine. 2011;**39**:1723-1730. DOI: 10.1177/0363546511404202

[24] Saris DB, Vanlauwe J, Victor J, Haspl M, Bohnsack M, Fortems Y, et al. Characterized chondrocyte implantation results in better structural repair when treating symptomatic cartilage defects of the knee in a randomized controlled trial versus microfracture. The American Journal of Sports Medicine. 2008;**36**:235-246. DOI: 10.1177/0363546507311095

[25] Caron MM, Emans PJ, Coolsen MM, Voss L, Surtel DA, Cremers A, et al. Redifferentiation of dedifferentiated human articular chondrocytes: Comparison of 2D and 3D cultures. Osteoarthritis and Cartilage. 2012;**20**:1170-1178. DOI: 10.1016/j.joca.2012.06.016

[26] Peterson L, Vasiliadis HS, Brittberg M, Lindahl A. Autologous chondrocyte implantation: A long-term follow-up. The American Journal of Sports

Medicine. 2010;**38**:1117-1124. DOI: 10.1177/0363546509357915

[27] Filardo G, Kon E, Di Martino A, Iacono F, Marcacci M. Arthroscopic second-generation autologous chondrocyte implantation: A prospective 7-year follow-up study. The American Journal of Sports Medicine. 2011;**39**:2153-2160. DOI: 10.1177/0363546511415658

[28] Basad E, Ishaque B, Bachmann G, Stürz H, Steinmeyer J. Matrix-induced autologous chondrocyte implantation versus microfracture in the treatment of cartilage defects of the knee: A 2-year randomised study. Knee Surgery, Sports Traumatology, Arthroscopy. 2010;**18**:519-527. DOI: 10.1007/s00167-009-1028-1

[29] Mithofer K, Minas T, Peterson L, Yeon H, Micheli LJ. Functional outcome of knee articular cartilage repair in adolescent athletes. The American Journal of Sports Medicine. 2005;**33**:1147-1153. DOI: 10.1177/0363546504274146

[30] Minas T, Von Keudell A, Bryant T, Gomoll AH. The John Insall award: A minimum 10-year outcome study of autologous chondrocyte implantation. Clinical Orthopaedics and Related Research. 2014;**472**:41-51. DOI: 10.1007/s11999-013-3146-9

[31] Richardson SM, Kalamegam G, Pushparaj PN, Matta C, Memic A, Khademhosseini A, et al. Mesenchymal stem cells in regenerative medicine: Focus on articular cartilage and intervertebral disc regeneration. Methods. 2015;**99**:69-80. DOI: 10.1016/j.ymeth.2015.09.015

[32] Vega A, Martin-Ferrero MA, Del Canto F, Alberca M, Garcia V, Munar A, et al. Treatment of knee osteoarthritis with allogeneic bone marrow mesenchymal stem cells: A randomized controlled trial. Transplantation. 2015;**99**(8):1681-1690. DOI: 10.1097/TP.0000000000000678

[33] Pak J, Lee JH, Lee SH. Regenerative repair of damaged meniscus with autologous adipose tissue-derived stem cells. BioMed Research International. 2014;**2014**:1-10. DOI: 10.1155/2014/436029

[34] Kang JY, Chung CW, Sung JH, Park BS, Choi JY, Lee SJ, et al. Novel porous matrix of hyaluronic acid for the three-dimensional culture of chondrocytes. International Journal of Pharmaceutics. 2009;**369**:114-120. DOI: 10.1016/j.ijpharm.2008.11.008

[35] Claros S, Rodríguez-Losada N, Cruz E, Guerado E, Becerra J, Andrades JA. Characterization of adult stem/progenitor cell populations from bone marrow in a three-dimensional collagen gel culture system. Cell Transplantation. 2012;**21**:2021-2032. DOI: 10.3727/096368912X636939

[36] Re'em T, Witte F, Willbold E, Ruvinov E, Cohen S. Simultaneous regeneration of articular cartilage and subchondral bone induced by spatially presented TGF-beta and BMP-4 in a bilayer affinity binding system. Acta Biomaterialia. 2012;**8**:3283-3293. DOI: 10.1016/j.actbio.2012.05.014

[37] Dhollander AA, Liekens K, Almqvist KF, Verdonk R, Lambrecht S, Elewaut D, et al. A pilot study of the use of an osteochondral scaffold plug for cartilage repair in the knee and how to deal with early clinical failures. Arthroscopy. 2012;**28**:225-233. DOI: 10.1016/j.arthro.2011.07.017

[38] Delcogliano M, de Caro F, Scaravella E, Ziveri G, DeBiase CF, Marotta D, et al. Use of innovative biomimetic scaffold in the treatment for large osteochondral lesions of the knee. Knee Surgery, Sports Traumatology, Arthroscopy. 2014;**22**:1260-1269. DOI: 10.1007/s00167-013-2717-3.

[39] Becher C, Ettinger M, Ezechieli M, Kaps C, Ewig M, Smith T. Repair of retropatellar cartilage defects in the knee with microfracture and a cell-free polymer-based implant. Archives of Orthopaedic and Trauma Surgery. 2015;**135**:1003-1010. DOI: 10.1007/s00402-015-2235-5

[40] Makadia HK, Siegel SJ. Poly lactic-co-glycolic acid (PLGA) as biodegradable controlled drug delivery carrier. Polymers. 2011;**3**:1377-1397. DOI: 10.3390/polym3031377

[41] Maudens P, Seemayer CA, Thauvin C, Gabay C, Jordan O, Allémann E. Nanocrystal-polymer particles: Extended delivery carriers for osteoarthritis treatment. Small. 2018;**14**:1-9. DOI: 10.1002/smll.201703108

[42] Lin JB, Poh S, Panitch A. Controlled release of anti-inflammatory peptides from reducible thermosensitive nanoparticles suppresses cartilage inflammation. Nanomedicine. 2016;**12**:2095-2100. DOI: 10.1016/j.nano.2016.05.010

[43] Natarajan V, Krithica N, Madhan B, Sehgal PK. Formulation and evaluation of quercetin polycaprolactone microspheres for the treatment of rheumatoid arthritis. Journal of Pharmaceutical Sciences. 2011;**100**: 195-205. DOI: 10.1002/jps.22266

[44] Nooeaid P, Salih V, Beier JP, Boccaccini AR. Osteochondral tissue engineering: Scaffolds, stem cells and applications. Journal of Cellular and Molecular Medicine. 2012;**16**:2247-2270. DOI: 10.1111/j.1582-4934.2012.01571.x

[45] Tamai N, Myoui A, Hirao M, Kaito T, Ochi T, Tanaka J, et al. A new biotechnology for articular cartilage repair: Subchondral implantation of a composite of interconnected porous hydroxyapatite, synthetic polymer (PLA-PEG), and bone morphogenetic protein-2 (rhBMP2). Osteoarthritis and Cartilage. 2005;**13**:405-417. DOI: 10.1016/j.joca.2004.12.014

[46] Miller RE, Miller RJ, Malfait AM. Osteoarthritis joint pain: The cytokine connection. Cytokine. 2014;**70**(2):185-193. DOI: 10.1016/j.cyto.2014.06.019

[47] Mease PJ, Hanna S, Frakes EP, Altman RD. Pain mechanisms in osteoarthritis: Understanding the role of central pain and current approaches to its treatment. The Journal of Rheumatology. 2011;**38**(8):1546-1551. DOI: 10.3899/jrheum.100759

[48] Inacio MC, Pratt NL, Roughead EE, Paxton EW, Graves SE. Opioid use after total hip arthroplasty surgery is associated with revision surgery. BMC Musculoskeletal Disorders. 2016;**17**:122. DOI: 10.1186/s12891-016-0970-6

[49] Altman R, Hochberg M, Gibofsky A, Jaros M, Young C. Efficacy and safety of low-dose SoluMatrix meloxicam in the treatment of osteoarthritis pain: A 12-week, phase 3 study. Current Medical Research and Opinion. 2015;**31**(12):2331-2343. DOI: 10.1185/03007995.2015.1112772

[50] McCarberg B. Tramadol extended-release in the management of chronic pain. Therapeutics and Clinical Risk Management. 2007;**33**:401-410

[51] Benke M, Shaffer B. Viscosupplementation treatment of arthritis pain. Current Pain and Headache Reports. 2009;**13**(6):440-446. DOI: 10.1007/s11916-009-0072-3

[52] Kirchner M, Marshall D. A double-blind randomized controlled trial comparing alternate forms of high molecular weight hyaluronan for the treatment of osteoarthritis of the knee. Osteoarthritis and Cartilage.

2006;**14**(2):154-162. DOI: 10.1016/j.joca.2005.09.003

[53] Mithoefer K, McAdams T, Williams RJ, Kreuz PC, Mandelbaum BR. Clinical efficacy of the microfracture technique for articular cartilage repair in the knee: An evidence-based systematic analysis. The American Journal of Sports Medicine. 2009;**37**:2053-2063. DOI: 10.1177/0363546508328414

[54] Gudas R, Gudaite A, Mickevicius T, Masiulis N, Simonaityte R, Cekanauskas E, et al. Comparison of osteochondral autologous transplantation, microfracture or debridement techniques in articular cartilage lesions associated with anterior cruciate ligament injury: A prospective study with a 3-year follow up. Arthroscopy. 2012;**29**:89-97. DOI: 10.1016/j.arthro.2012.06.009

[55] Gomoll AH. Microfracture and augments. The Journal of Knee Surgery. 2012;**25**:9-15. DOI: 10.1055/s-0031-1299654

[56] Knutsen G, Engebretsen L, Ludvigsen TC, Drogset JO, Grontvedt T, Solheim E, et al. Autologous chondrocyte implantation compared with microfracture in the knee: A randomized trial. The Journal of Bone and Joint Surgery. American Volume. 2004;**86**:455-464. DOI: 10.1.1.555.8837

[57] Bennell KL, Hinman RS. A review of the clinical evidence for exercise in osteoarthritis of the hip and knee. Journal of Science and Medicine in Sport. 2011;**14**:4-9. DOI: 10.1016/j.jsams.2010.08.002

[58] Pichler K, Loreto C, Leonardi R, Reuber T, Weinberg AM, Musumeci G. RANKL is downregulated in bone cells by physical activity (treadmill and vibration stimulation training) in rat with glucocorticoid-induced osteoporosis. Histology and Histopathology. 2013;**28**:1185-1196. DOI: 10.14670/HH-28.1185

[59] Musumeci G, Loreto C, Leonardi R, Castorina S, Giunta S, Carnazza ML, et al. The effects of physical activity on apoptosis and lubricin expression in articular cartilage in rats with glucocorticoid-induced osteoporosis. Journal of Bone and Mineral Metabolism. 2013;**31**:274-284. DOI: 10.1007/s00774-012-0414-9

[60] Andrade R, Vasta S, Pereira R, Pereira H, Papalia R, Karahan M, et al. Knee donor-site morbidity after mosaicplasty – A systematic review. Journal of Experimental Orthopaedics. 2016;**3**(1):31. DOI: 10.1186/s40634-016-0066-0

[61] Shin YH, Kim MH, Ko JS, Park JA. The safety of simultaneous bilateral versus unilateral total knee arthroplasty: The experience in a Korean hospital. Singapore Medical Journal. 2010:51: 44-49. PMID: 20200775

[62] Krishnamoorthy S, Govindaraj AB, Vivek AN, VijayKumar S, Anand M. Functional outcome of single stage bilateral total knee replacement measured using oxford knee score. Journal of Arthroscopy and Joint Surgery. 2018. DOI: 10.1016/j.jajs.2019.01.010

Chapter 2

Epigenetics and Cartilage Regeneration

Samina Hyder Haq, Iqraa Haq, Atheer Ali Alsayah, Abir Alamro and Amani AlGhamedi

Abstract

Regenerative cartilage therapy has great potential for the treatment of debilitating diseases such as osteoarthritis and rheumatoid arthritis. Recent advances in the field of epigenetics have enabled us to understand more clearly the role of micro RNAs, DNA methylations and histone modification in disease progression, as well as its potential role in disease prevention. However, a thorough understanding of the external dietary and environmental factors that could affect the epigenetic events could be the key to unravelling novel therapeutic strategies for these diseases. There is, therefore, a need for identifying certain dietary or environmental factors that could change this downward epigenetics signalling cascade, stop or retard cartilage degradation and promote cartilage regeneration.

Keywords: cartilage regeneration, DNA methylations, epigenetics, therapeutic dietary supplements, DNMT inhibitors

1. Introduction

Articular cartilage is an aneural, avascular, alymphatic specialized fibrous connective tissue which covers the articulating surface of synovial joints. This is characterized by a small number of morphologically distinct populations of chondrocytes, which are primarily responsible for production, organization and maintaining the extensive network of an extracellular matrix. The balance between the hydration of matrix proteoglycans (PGs) and the resistance offered by the extensive network of the fibrous structure of the collagen provides the hydrodynamic load-bearing properties of articular cartilage, which is critical for joint movements and smooth transmission of mechanical compression across the joint. As articular cartilage is originally derived from the hyaline cartilage template, it is also classified as permanent hyaline cartilage. After the original phase of cartilage production, differentiation and resorption and closure of growth plate cartilage at puberty, it remains as a part of bone throughout the adult life. It is divided into four distinct horizontal layers: the superficial, transitional, deep and calcified cartilage layers (**Figure 1**).

The thin superficial zone protects the deeper layers from shear stress and injury and makes up 10–20% of articular cartilage thickness. This layer is characterized by small flattened disc-shaped chondrocytes, comparatively low proteoglycan content and densely packed layers of uniformly formed collagen fibres, which gives the characteristic hyaline opacity to cartilage. This layer is in direct contact to synovial fluids and is responsible for most of the tensile strength of the cartilage as well as

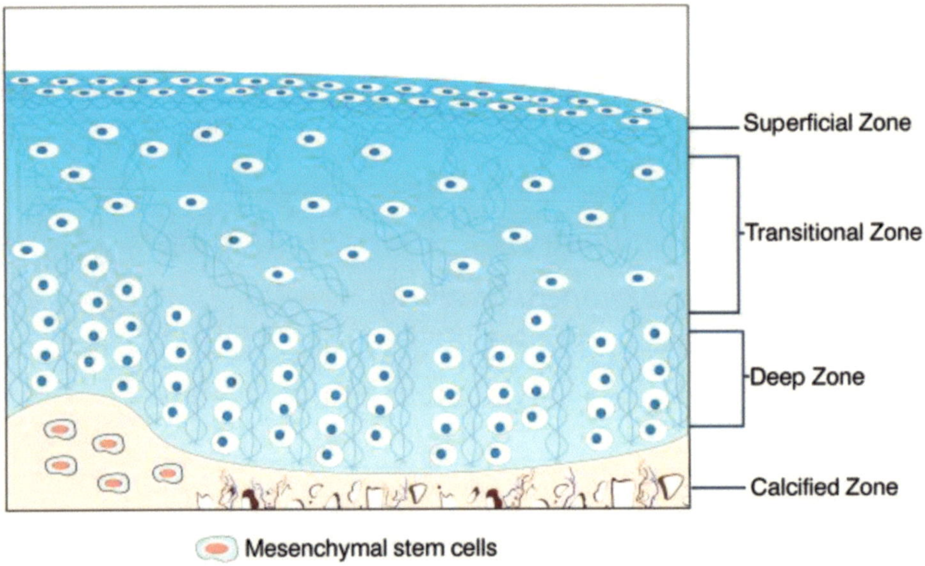

Figure 1.
Stratification of articular cartilage.

takes the direct brunt of inflammatory cytokines. It is well documented that the chronic inflammation in joints in osteoarthritis (OA) patients is due to synovial macrophages and high inflammatory cytokines that initiate the aggregenase, MMPs and other destructive enzymes. Immediately below the superficial zone is the middle or transitional layer which provides the functional bridge between the superficial and deep layers. The middle layer comprises of 40–60% of the total cartilage volume. In this layer, the chondrocytes attain a more rounded or spherical shape, the contents of proteoglycans increase, and thicker collagen fibres provide an oblique transitional network intermediate between the tangential superficial and radial deep layers. The deep layer is characterized by relatively mature rounded chondrocytes arranged in longitudinal columns, high proteoglycan contents, the largest diameter collagen fibrils in a radial disposition and the lowest water concentration. This zone represents approximately 30% of the total cartilage volume. The calcified layer is characterized by rounded hypertrophic chondrocytes surrounded by large clear lacunae. This is the area where the chondrocytes reach their terminal hypertrophic stage and the cartilage is ultimately being replaced by bone.

2. Molecular heterogeneity of articular cartilage

The extracellular phase of cartilage, and all connective tissues, consists of collagen fibres and a polysaccharide-rich ground substance. The polysaccharide constituents have been characterized as proteoglycans containing chains of chondroitin 4 sulphate, chondroitin 6 sulphate and keratin sulphate covalently linked to a central core protein [1].

2.1 Types of collagen present in cartilage

Articular cartilage consists of type II collagen as the major fibril-forming collagen, accompanied by lesser quantities of minor collagen which provide the tensile strength and help in maintaining the fine balance of the extracellular matrix. However, little is known about the processing of these minor collagens and their role in the progression of cartilage degeneration and regeneration. Minor collagens found in articular cartilage along with type II collagen are type VI, IX, X, XI, XII and XIV.

Type VI collagen constitutes only 1–2% of the total collagen in adult articular cartilage and it is mainly rich in the pere-cellular matrix and involved in the integration and attachment of chondrocytes [2]. In articular cartilage, chondrocytes in the middle and deep layers are embedded in pere-cellular matrix enriched with a high content of proteoglycans and hyaluronic acid. Increased levels of type VI collagen are found in the experimental model of osteoarthritis (OA) and human OA [3]. Higher levels of type VI collagen found in OA emphasizes its role as a bridge between the extracellular matrix and the chondrocyte surface, thus influencing the signalling pathways from the extracellular matrix into the cells [4].

Type IX collagen makes up 1–5% of the total collagen in adult articular cartilage and 10% in foetal cartilage [5]. It is usually present in close association with type II collagen found in growth plate cartilage and adult articular cartilage [6]. Type IX collagen is extensively crosslinked to type II collagen through oxidation of lysyl residue bonds forming a unique hetero-fibrillar structure [7]. Type IX collagen is crucial for the maintenance of cartilage matrix and formation of a collagen fibril meshwork. Decreased expression of type IX collagen in the cartilage was thought to render the matrix more prone to mechanical forces and degradation, resulting in the pathogenesis of OA [8].

Type X constitutes about 1% of the total collagen found in articular cartilage [9]. It was revealed that 45% of the total collagen produced by the hypertrophic chondrocytes is type X collagen [10]. Type X collagen, as produced exclusively by hypertrophic chondrocytes, indicated its unique role in mineralization. The hypertrophic chondrocytes synthesized a variety of proteins and enzymes which help in the transition of extracellular matrix from cartilage to bone. Apart from type X collagen, hypertrophic chondrocytes also synthesize a variety of matrix metalloproteinases as well as alkaline phosphatase enzymes, which are not usually secreted by the normal proliferating chondrocytes. As type X collagen has a direct role in mineralization, it has been found to be expressed in human OA especially in the vicinity of lesions, but not in the healthy human articular cartilage [11].

Type XI collagen constitutes 3–10 % the total adult articular and foetal cartilage, respectively [2]. Type XI collagen is normally crosslinked to each other in cartilage, this crosslinking results in the formation of mature type XI collagen with the help of type II and type IX collagen. It has been shown that a mutation in type XI collagen caused an increase in degradation of type II collagen in articular cartilage [12]. Lu et al. observed that immunization of rats with homologous type XI collagen led to chronic and relapsing arthritis with different genetics and joint pathology than arthritis induced with homologous type II collagen [12]. The role of type XI collagen in cartilage collagen fibril formation and assembly is not clear; type XI collagen may regulate cartilage formation and it was the first collagen deposited by mesenchymal stem cells undergoing chondrogenic differentiation [13]. Type XII shares structural homologies with type IX and type XIV collagen [14]. Type XII collagen is implicated in fibril formation, cell adhesion, fibrosis and osteogenesis, and in areas of high mechanical stress, it may serve as a protector of tissue integrity [15]. Type XII collagen is associated with articular cartilage and growth plate cartilage during rat forelimb development and may be important for microenvironment that supports the hyaline cartilage formation [16].

Type XIV collagen is a large nonfibrillar extracellular matrix protein structurally similar to type XII collagen. In cartilage, a population of type XIV exists as chondroitin sulphate proteoglycans (PGs) as it is sensitive to chondroitinase ABC and AC treatments [17]. Its association with other cartilage collagens such as type I, II,

V and VI are reported, but it also interacts with heparin CD44 and cartilage oligomeric matrix protein [18]. It is found in areas of high mechanical stress similar to type XII collagen, suggesting its role in fibrillogenesis and maintaining the integrity and mechanical property of the tissue.

2.2 Types of PGs in different layers

Proteoglycans have the highest concentrations in the intermediate zone and lowest in the superficial and deep zones. Small PGs comprise of less than 10% of the total PG content in the cartilage matrix. Most are aggrecans (large PGs) with approximately 150 GAG chains (chondroitin sulphate and keratin sulphate and both O-linked and N-linked oligosaccharides attached).The GAGs are heterogeneously distributed along the protein core, with CS-rich and KS-rich regions, respectively. The protein core itself is heterogeneous with three globular regions. Aggrecan varies significantly in length, molecular weight and composition with the amount of KS-rich molecules and ratios of chondroitin 6-sulphate and chondroitin 4-sulphate increasing throughout development and ageing. Most aggrecans in cartilage are attached to a hyaluronic (HA) molecule via a globular (HABR) region; this binding was stabilized by a link protein. Several hundred aggrecans are attached to a single HA core molecule, the latter being a non-sulphated disaccharide chain up to 4 μm in length. PGs are closely associated with collagen fibrils and are thought to be involved in their structural organization and maintaining their compressive stiffness.

There is now conclusive evidence of the fact that OA is not simply due to wear and tear and a result of ageing; but in numerous studies, it has been reported that early onset of OA is due to activation of inflammatory response. These inflammatory responses could be due to increased oxidative stress to the tissues, resulting in initiation of catabolic enzymes and factor that actively breakdown the major extracellular matrix components of cartilage, namely type II collagen, and the proteoglycans and aggrecan.

3. Control of chondrogenesis

The commitment of mesenchymal cells to the chondrogenic lineage is the key event in bone formation. Work over the past few decades, using both in vivo and in vitro systems, has identified a number of signalling and transcription factors as well as cell shape that regulates the progressive change in chondrocyte phenotype, from their initial induction to their terminal fate. The disruption of these finely tuned pathways for chondrocyte maturation can result in skeletal pathology. A thorough knowledge of these signalling pathways would help us to identify the factors that maintain chondrocyte proliferation and differentiation. Some of the major signalling pathways are described below.

3.1 Bone morphogenic protein signalling

Bone morphogenic proteins (BMPs) are identified as positive regulators of chondrogenesis and endochondral ossification. BMPs are a member of the transforming growth factor beta (TGβ) superfamily that has wide-ranging biological activity, ranging from cellular regulation of proliferation, apoptosis, differentiation and migration [19, 20]. BMP signalling is mediated by their receptors BMPR1a, BMPR1b and BMPR2, leading to the SMAD signalling pathway [19]. In cartilage, it initiates cartilage synthesis and decreases the activity of catabolic cytokines such as IL-1, IL-6, IL-8, MM1 and MM13 [21, 22]. Though there are several members of

Bone morphogenic protein (BMP) growth factors, most promising among them in the treatment of OA is BMP-7, which promotes the cartilage-specific extracellular matrix proteins such as collagen II and VI, decorin, fibronectin and hyaluronate (HA) by upregulation of hyaluronan synthase [23, 24]. In experiments when it was applied to other types of cells in knee, BMP-7 has shown to increase Extracellular matrix (ECM) in synovial and bone marrow-derived Mesenchymal Stem cell (MSC), both alone and in combination with TGFβ [25]. This profound anabolic effect of BMP-7 is due to its regulatory properties of modulating other growth factors such as insulin-like growth factor 1(ILGF1 and fibroblast growth factor (FGF)) [26]. Despite having anabolic activity, BMP-7 has not shown to induce chondrocyte hypertrophy or other changes in the chondrocyte phenotype, nor did BMP-7-treated animal knee display any proliferation of fibroblast or osteocyte [25]. These properties make it a promising therapy for OA.

3.2 Transforming growth factor (TGF) signalling pathway

TGFβ is a cytokine secreted by many cells; it plays an important role in cell proliferation, differentiation, development, apoptosis, tissue homeostasis and the immune system. Signalling occurs through SMAD pathways. TGβ1 is shown to be involved in chondrocyte proliferation and remarkable reduction of catabolic activity of IL1 and TNF [27]. Studies have shown a significant enhancement of cartilage repair with the application of TGF-β1 in scaffold applied to defect, and in human MSC transfected with TGF-β1 gene via an adenovirus [28, 29]. Numerous human trials are underway for the treatment of different stages of OA with the injections of TGF-β1 in the knee, which showed TGF-β1 as a promising therapy.

3.3 Fibroblast growth factor signalling pathway

Fibroblast growth factor (FGF) family plays an important role in human embryonic development, cell growth, morphogenesis, tissue repair, tumour growth and invasion. FGFs are heparin-binding proteins and interact with heparan sulphate proteoglycans on the cell surface for signal transduction. Vincent et al. proposed that in articular cartilage, the chondrocytes are surrounded by a pool of FGF-2. This mediated the chondrocyte activation on cartilage loading and release of FGF-2 in response to injury. They proposed that FGF-2 antagonizes the PG degradation by IL-1 or other catabolic stimuli, thus it has an anti-catabolic chondroprotective role [30]. However, the role of FGF-2 in the production of ECM is controversial and its role as pro-catabolic or anti-catabolic is debatable. Furthermore, FGF-2 has been shown to suppress type II collagen and PG synthesis and promote the expression of aggregenase and TNF-α receptors [31, 32]. FGF-18 signalling through FGFR3 promotes chondrocyte proliferation at embryonic stages. When development is complete, the same receptor signalling suppresses chondrocyte proliferation and prevents chondrocyte differentiation hypertrophy [33, 34]. FGF-18 has also shown to exhibit the ability to stimulate type II collagen and PG synthesis, which makes it a promising therapy for OA.

3.4 Connective tissue growth factor

Connective tissue growth factor (CTGF) is an ECM-associated heparin-binding protein, which plays an important role in cellular proliferation, migration, adhesion, survival and synthesis of ECM proteins. CTGF has shown to play an important role in skeletal tissue and initial chondrocyte proliferation and differentiation in growth plate cartilage [35]. Nishida and colleagues demonstrated that local administration of recombinant CTGF gelatin hydrogel stimulated cartilage repair in rat model [36]. Other studies showed that the bone marrow-derived mesenchymal stem cells when transfected with CTGF provided hyaline-like cartilage regeneration similar to normal cartilage in a rabbit model of focal articular cartilage defects [37]. However, further studies are needed to elucidate the critical role of CTGF for the protection and regeneration of cartilage.

3.5 Insulin-like growth factor (IGF)

IGF-I and IGF-II both were reported to control the cartilage destruction [38]. IGF-I is a known anabolic factor for chondrocytes and thought to regulate the skeletal development in the embryo [39]. Although IGF-I has been reported as being involved in chondrocyte proliferation and maturation, its exact role in OA has not been clearly known as it was found that the expression of IGF-I was upregulated rather than downregulated in synovial fluids and in articular cartilage [40]. However, the role of IGF-II in combating inflammatory response in OA was found to be more promising and ideal for cartilage regeneration. It has been reported that in the presence of IL-1β, IGF-II significantly inhibited MMP expression and promoted cartilage production in normal human chondrocytes. IGF-II has also shown to have a similar effect on OA chondrocyte, which expresses high levels of IL-1β mRNA [41]. The role of IGF-II was reported to be more chondroprotective and maintaining the extracellular matrix and preventing its destruction in adverse conditions.

4. Cell signalling in chondrogenesis

Gene expression changes during different stages of endochondral ossification. The immature chondrocytes in the resting zone express the transcription factors Sox 5, Sox6, Sox9 and the structural protein type II collagen and aggrecan. The pre-hypertrophic zone is characterized by the presence of parathyroid receptor 1(PTH-1R) and Indian hedgehog expression (Ihh). The next stage goes into the early hypertrophic zone, which is characterized by type X collagen and alkaline phosphatase enzyme expression, and, subsequently, the reduced amount of type II collagen and reduced expression of Sox5, Sox6 and Sox9 transcription factors. Finally, the chondrocytes proceed to their final phase of a late hypertrophic stage, which is characterized by the expression of vascular endothelial growth factor A (VEGFA), matrix metalloproteinase 13(MM13) and osteopontin. These changes in gene expression herald the cartilage ECM being replaced by bone.

Wnt signalling is important for many developmental processes. It has been shown that activation of Wnt signalling promotes osteoblast differentiation but inhibits chondrocyte differentiation of MSC [42, 43]. Wnt signalling acts through β-catenin to promote chondrocyte hypertrophy and reports suggested that genetic inactivation of β-catenin increased Sox9 expression both in the intramembranous bone formation and endochondral ossification [44, 45]. It was also reported that osteoblast precursor lacking β-catenin expression can develop into chondrocytes [42]. Wnt signalling is also important for proper orientation of chondrocyte column in growth plate cartilage.

Ihh signalling is a key regulator of pre-hypertrophic and early hypertrophic chondrocytes. Ihh signalling directly affects chondrocyte proliferation, premature chondrocyte hypertrophy and failure of osteoblast development and endochondral bone.

Runx2 and Runx3 are members of the Runx transcriptional factors family important for chondrocyte hypertrophy. Several studies demonstrated that ectopic expression of Runx2 in immature chondrocytes leads to the expression of hypertrophic markers such as COLX α1, MM13 and VEGF [46–48].

As cartilage is an avascular tissue and its nutritional needs are met by surrounding synovial fluids, chondrocytes are adapted to survive in low oxygen levels and they secrete hypoxia-inducible factor 1 (HIF-1) which insures its survival and maintenance in low oxygen tension. Synthesis of type II collagen and aggrecan is upregulated in low oxygen levels, and also, it is associated with the rounded chondrocytic morphology. In the presence of high oxygen tension, chondrocytes become more spindle shaped. HF-1 also showed inhibition of type I collagen synthesis by inhibiting its promoter activity [49].

5. Epigenetic control of chondrogenesis

In the growth of long bone formation, the chondrocyte passes through discrete stages of proliferation, maturation, hypertrophy, calcification and apoptosis, so it offers a very good model of cellular differentiation and ageing. The detailed underlying molecular mechanisms that drive these changes are still not fully known, but epigenetic modifications are thought to play a pivotal role in the differentiation of chondrocytes. Epigenetic changes include DNA methylation, histone modification and microRNAs (miRNAs).

5.1 DNA methylation

DNA methylation involves the addition of a methyl group to a DNA at CpG dinucleotide, to convert cytosine to 5-methylcytosine. CpG islands are usually clustered near the promoter in about 30% of the gene. Methylation of these sequences results in silencing of these genes, and vice versa, hypo-methylation results in expression of the respective genes. DNA methylation factors are established and modified according to the environmental factors by three DNA methyltransferases (DNMT1, DNMT3a and DNMT3b). Earlier studies using chick embryos indicate the possible role of methylation in gene expression of type I and type II collagen in chondrocyte differentiation and dedifferentiation [50]. In our studies on chick chondrocytes in culture, we noticed a strong correlation of chondrocyte morphology to DNA methylation status as shown in **Figure 2**. The chondrocytes when treated with DNMT inhibitor 5-aza-2'deoxycytidine exhibit fibroblastic morphology and express type I and type X collagen with an upregulation of alkaline phosphatase enzyme [51]. Two CpG sites within the type X collagen promoter appear to be demethylated during MSC differentiation into chondrocyte morphology [52]. Recently, it was demonstrated that Wnt signalling caused both repressive chromatin mark (H3K27me3) and DNA methylation over the SOX9 promoter and that Wnt-induced irreversible silencing of Sox9 gene requires DNA methylation of this locus that is specifically countered by FGF signalling [53]. FGF blocks the recruitment of DNMT3a to the SOX-9 promoter by inducing the interaction and phosphorylation of DNMT3a by extracellular kinases ERK 1and ERK 2. Similarly, a number of studies indicated the control of Runx2 promoter activation by methylation. The number of MMP promoters show decreased methylation at single CpG island in OA cartilage as compared to normal.

5.2 Histone modifications

Gene regulation is also controlled through the close packaging of eukaryotic DNA into nucleosomes. Nucleosomes are thought to be repressive for

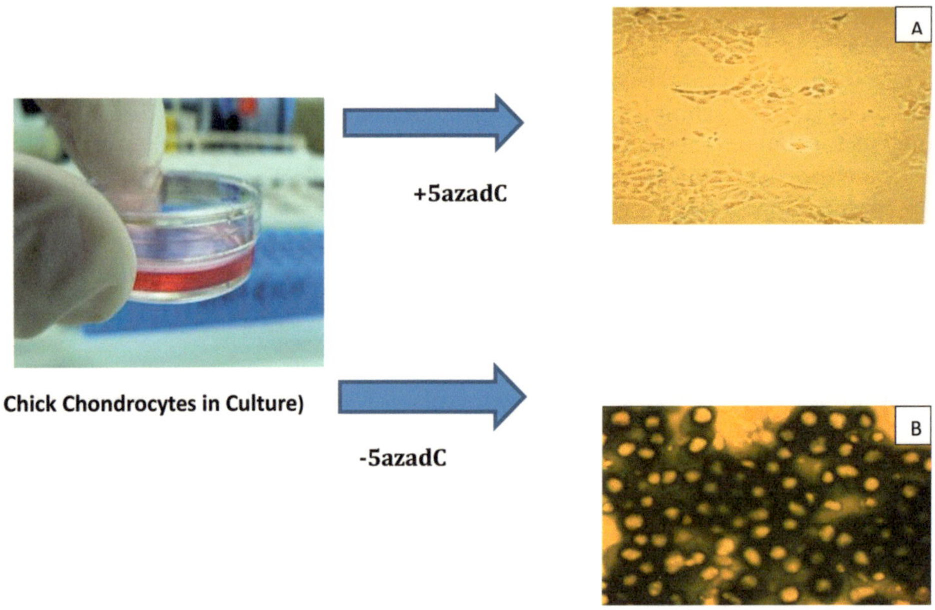

Figure 2.
The effect of culture conditions on the morphology of chondrocytes: When chick chondrocytes from caudal region sternum were grown in the presence of demethylation drug 5aza-2'deoxyctydine (5azadC), (A) the chondrocytes assume more flattened fibroblastic morphology and show no staining with alcian blue (stain specific for sulphate PG). However, the control chondrocyte without any treatment showed extensive ECM staining (B).

transcription; but through the post-translational modification of histones such as acetylation, phosphorylation, methylation and ubiquitination, this inhibition can be regulated.

Acetylation is mediated through acetyltransferase (HAT) and occurs on specific lysine residues on the N-terminal tails of histones, loosening the histones: DNA interactions, thus employing the access of transcriptional factors to the DNA. Deacetylation is of two types, one that requires Zn-catalysed deacetylation (HDAC) and the sirtuin deacetylase that requires NAD+, and removes these acetyl groups resulting in hypo-acetylation. Numerous transcriptional activators or repressors recruit HDAC and HAT activity.

Histone methylation is important for the formation of active and inactive genomic regions and is associated with transcription activation and silencing. Methylation of histone tails of lysine and arginine residues is catalysed by histone methyltransferase (HMT) and protein arginine methyltransferase (PRMT) which can add one or more methyl groups to regulate transcription [54]. Although histone methylation is more dynamic than DNA methylation, some specific histone methylation is tightly regulated and maintained through DNA replication. HDAC can block cytokine-induced PG release and cartilage resorption in cartilage explant model indicating that HDAC activity is important for the catabolic activity of chondrocytes [55, 56].

5.3 Micro RNA

MiRNA is a small 20–23 base pair-long cytoplasmic RNA that regulates post-transcriptional gene expression through binding to target mRNA. This interaction of miRNA with the target mRNA results in degradation of mRNA, thus suppression of translation. The first studied miRNA in cartilage was miR-140, which was first identified as cartilage restricted in developing zebrafish [57]. In humans, the expression of miR-140 increases during chondrogenesis and is more abundant in articular cartilage, but its expression is reduced in OA [58]. It has also been reported that the expression of miR-140 is regulated by the cartilage-specific master transcriptional factor Sox-9 in zebrafish and mammalian cells [59].

6. Epigenetics as a future therapy for cartilage regeneration

Articular cartilage has a relatively high incidence of damage due to several factors such as injury, trauma and inflammation. The inflammatory markers could induce a number of MMPs, which could degrade the ECM, as the cartilage has a limited ability to repair and regenerate, resulting in a total loss of cartilage tissue. The destruction and loss of articular cartilage is also central to the development of OA. The research work over the past few decades confirms that epigenetics plays a pivotal role in the phenotypic modulation that articular chondrocytes undergo during OA. Epigenetics changes the normal chondrocytes to 'altered' chondrocytes that overexpress the cartilage-degrading proteins or enzymes such as collagenases and aggregenase and inflammatory mediators. This disruption in homeostatic balance between the matrix production and ECM destruction results in the progression of OA. There is a direct pathological loop that involves inflammation and epigenetic modifications, which accelerates disease progression. Until now, no detailed global methylation analysis has been performed in the pathogenesis of OA. Low penetrance polymorphism in the population partly due to epigenetic modification is the reason for limited data generation to aid in the identification of genes responsible for the genetic susceptibility to OA. A number of inflammatory genes have been identified which are controlled through epigenetics and are directly involved in the pathogenesis of OA (**Table 1**).

6.1 Future prospects in cartilage regeneration

MCS is becoming a more popular source of cells for cartilage regeneration due to in vitro expansion without running the risk of losing their phenotype. However, MSC tends to develop hypertrophic phenotype and further differentiation into the endochondral bone formation. It is becoming more crucial to carefully examine the detailed molecular and epigenetic events that lead the transformation of a chondrocyte to its terminally differentiated pathway. There is a growing need to develop strategies to control chondrocyte hypertrophy and be able to arrest the chondrocyte at one desirable phenotypic stage that helps to maintain the cartilage-specific ECM as described in **Figure 3**. With the current epigenetic knowledge, it is possible to identify a number of epigenetic factors as listed in **Table 1** that can make cartilage regeneration possible.

Other option in cartilage regeneration is the application of hydrogel through injection or through arthroscopy. These hydrogels are capable of controlled release of chondroinductive and chondroprotective drugs [60–62]. These cell-laden hydrogels can be combined with other types of solid scaffolds such as collagen sponge, decellularized cartilage as well as synthetic scaffolds of polyglycolic acid for cartilage repair and clinical applications.

Chondrocyte stage	Marker	Function gene	Epigenetic regulation	References
Superficial zone	Col2a1	Cartilage specific	miRNA, Histone modification	[1]
	Col6a1	Pere-cellular chondrocyte	DNA methylation	[2]
	Col9a1	Cartilage specific	DNA methylation	[3]
	ACAN	Cartilage specific	miRNA, Histone modification	[4, 5]
	HIF1α, HIF2α	Chondrocyte viability	miRNA	[6]
Transitional	IGFII	Chondrocyte proliferation and integrity		[7]
	SOX-9	Chondrocyte differentiation	DNA methylation, miRNA, histone methylation	
	T3 +PTH	Cartilage tissue regeneration	Histone modification	[8, 9]
	NFAT	Cartilage matrix	Histone methylation	[10]
	FGFR3	Chondrogenesis	DNA methylation	[10]
	TGFβ1-β3	Chondrocyte proliferation	DNA methylation	[11, 12]
	BMP-7	Cartilage specific ECM	Histone modification, DNA methylation	[11, 13]
Deep/ Calcifying	Col10a1	Chondrocyte hypertrophy	DNA methylation	[14]
	Col1α1	Bone formation	DNA methylation	[15]
	Osteocalcin	In calcification	DNA methylation	[16]
	Osteopontin	Bone formation	DNA methylation	[16]
	ALPL	Chondrocyte hypertrophy	DNA methylation	[17]
	RUNX 2	Chondrocyte hypertrophy	DNA methylation, miRNA	[18, 19]
	ADAM	Cartilage remodelling	DNA methylation, miRNA	[10]
	IHH	Cartilage hypertrophy	DNA methylation	[10]
	TGFβ2	Hypertrophy	DNA methylation	[20, 21]
	MMP13	Cartilage remodelling	DNA methylation, histone modification, miRNA	[1]
OA cartilage	HDAC	Up regulated in OA	Histone modification	[22]
	IL-1β	Inflammation	DNA methylation, miRNA	[23]
	TNFα	Inflammatory mediator	DNA methylation, miRNA, histone modification	[11]
	MMP3	Up regulated in OA	DNA methylation	[24]
	MMP9	Up regulated in OA	DNA methylation	[24]
	ADAMS4	Expressed in OA	DNA methylation	[24]

Table 1.
Major Epigenetic events remodelling the regeneration of Cartilage.

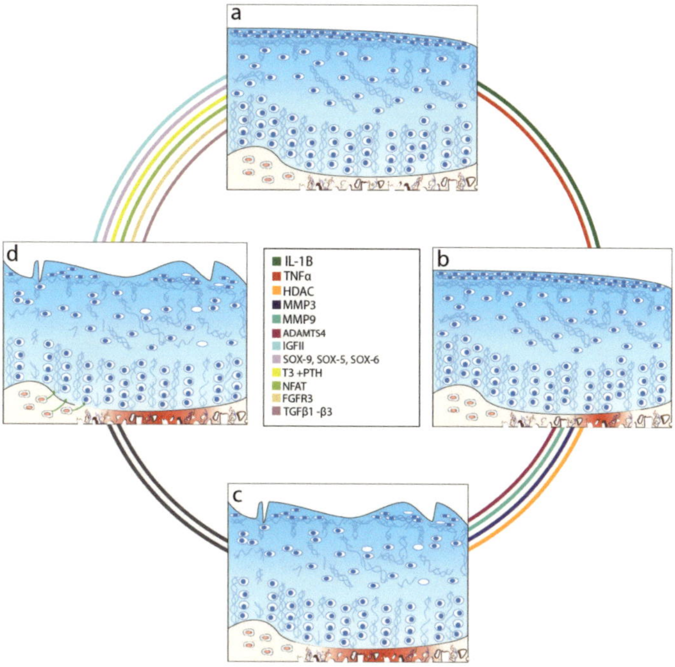

Figure 3.
*The role of epigenetics in cartilage degradation and regeneration. (a) Healthy articular cartilage with distinct stratifications. (b) As a result of high inflammatory markers such as IL-1β and TNFα, cartilage degradation takes place, with upregulation of a number of cartilage-degrading enzymes (e.g., HDAC, MMP3, MMP9, and ADAMTS4). (c) The onset of OA, which can be reversed with the help of MSC therapy and their initiations as shown in (d). The maintenance of healthy articular cartilage is achieved through a cascade of genes and their products, such as IGFII, SOX5, SOX6, SOX9, NFAT, FGFR3 and TGFβ1-β2. They are all controlled through epigenetics (***Table 1***). Future cartilage regeneration technique should involve the promotion of invasion and migration of MCSs to the lesion area and through various epigenetic signalling undergoing chondrogenesis and maintaining the cartilage.*

7. Conclusion

In summary, although there has been progress made in identifying factors outlining OA disease progression, a more detailed analysis of the factors surrounding the epigenetics should be conducted in order to reveal any potential therapies. The control of chondrogenesis via bone morphogenic protein signalling, transforming growth signalling, fibroblast growth factor signalling, connective tissue growth factor and insulin-like growth factor all play important roles in chondrocyte formation and destruction. This in addition to the fact that cellular mechanisms controlled by gene expression and epigenetic changes including DNA methylation, histone modification, and microRNAs can all help us gain an understanding of regenerative cartilage therapies.

Acknowledgements

The authors acknowledge the support and funding from the Deanship of Scientific Research, King Saud University, Riyadh.

Conflict of interest

There is no conflict of interest to declare.

Author details

Samina Hyder Haq[1*], Iqraa Haq[2], Atheer Ali Alsayah[1], Abir Alamro[1] and Amani AlGhamedi[1]

1 Department of Biochemistry, College of Science, King Saud University, Riyadh, Saudi Arabia

2 College of Medicine, Imperial College, London, UK

*Address all correspondence to: shaq@ksu.edu.sa

References

[1] McDevitt CA. Biochemistry of articular cartilage. Nature of proteoglycans and collagen of articular cartilage and their role in ageing and in osteoarthrosis. Annals of the Rheumatic Diseases. 1973;**32**(4):364-378

[2] Eyre D. Collagen of articular cartilage. Arthritis Research. 2002;**4**(1):30-35

[3] Pullig O, Weseloh G, Swoboda B. Expression of type VI collagen in normal and osteoarthritic human cartilage. Osteoarthritis and Cartilage. 1999;7(2):191-202

[4] Horikawa O, Nakajima H, Kikuchi T, Ichimura S, Yamada H, Fujikawa K, et al. Distribution of type VI collagen in chondrocyte microenvironment: Study of chondrons isolated from human normal and degenerative articular cartilage and cultured chondrocytes. Journal of Orthopaedic Science. 2004;**9**(1):29-36

[5] Eyre DR, Weis MA, Wu JJ. Articular cartilage collagen: an irreplaceable framework? European Cells & Materials. 2006;**12**:57-63

[6] Eyre DR, Apon S, Wu JJ, Ericsson LH, Walsh KA. Collagen type IX: evidence for covalent linkages to type II collagen in cartilage. FEBS Letters. 1987;**220**(2):337-341

[7] Wu JJ, Woods PE, Eyre DR. Identification of cross-linking sites in bovine cartilage type IX collagen reveals an antiparallel type II-type IX molecular relationship and type IX to type IX bonding. The Journal of Biological Chemistry. 1992;**267**(32):23007-23014

[8] Alizadeh BZ, Njajou OT, Bijkerk C, Meulenbelt I, Wildt SC, Hofman A, et al. Evidence for a role of the genomic region of the gene encoding for the alpha1 chain of type IX collagen (COL9A1) in hip osteoarthritis: A population-based study. Arthritis and Rheumatism. 2005;**52**(5):1437-1442

[9] Eyre DR, Wu JJ, Woods PE, Weis MA. The cartilage collagens and joint degeneration. British Journal of Rheumatology. 1991; **30**(Suppl 1):10-15

[10] Shen G. The role of type X collagen in facilitating and regulating endochondral ossification of articular cartilage. Orthodontics & Craniofacial Research. 2005;**8**(1):11-17

[11] Brew CJ, Clegg PD, Boot-Handford RP, Andrew JG, Hardingham T. Gene expression in human chondrocytes in late osteoarthritis is changed in both fibrillated and intact cartilage without evidence of generalised chondrocyte hypertrophy. Annals of the Rheumatic Diseases. 2010;**69**(1):234-240

[12] Lu S, Carlsen S, Hansson AS, Holmdahl R. Immunization of rats with homologous type XI collagen leads to chronic and relapsing arthritis with different genetics and joint pathology than arthritis induced with homologous type II collagen. Journal of Autoimmunity. 2002;**18**(3):199-211

[13] Xu J, Wang W, Ludeman M, Cheng K, Hayami T, Lotz JC, et al. Chondrogenic differentiation of human mesenchymal stem cells in three-dimensional alginate gels. Tissue Engineering. Part A. 2008;**14**(5):667-680

[14] Yamagata M, Yamada KM, Yamada SS, Shinomura T, Tanaka H, Nishida Y, et al. The complete primary structure of type XII collagen shows a chimeric molecule with reiterated fibronectin type III motifs, von Willebrand factor A motifs, a domain homologous to a noncollagenous region of type IX collagen, and short

collagenous domains with an Arg-Gly-Asp site. The Journal of Cell Biology. 1991;**115**(1):209-221

[15] Chiquet M, Birk DE, Bonnemann CG, Koch M. Collagen XII: Protecting bone and muscle integrity by organizing collagen fibrils. The International Journal of Biochemistry & Cell Biology. 2014;**53**:51-54

[16] Taylor DW, Ahmed N, Parreno J, Lunstrum GP, Gross AE, Diamandis EP, et al. Collagen type XII and versican are present in the early stages of cartilage tissue formation by both redifferentating passaged and primary chondrocytes. Tissue Engineering. Part A. 2015;**21**(3-4):683-693

[17] Watt SL, Lunstrum GP, McDonough AM, Keene DR, Burgeson RE, Morris NP. Characterization of collagen types XII and XIV from fetal bovine cartilage. The Journal of Biological Chemistry. 1992;**267**(28):20093-20099

[18] Giry-Lozinguez C, Aubert-Foucher E, Penin F, Deleage G, Dublet B, van der Rest M. Identification and characterization of a heparin binding site within the NC1 domain of chicken collagen XIV. Matrix Biology. 1998;**17**(2):145-149

[19] Goumans MJ, Mummery C. Functional analysis of the TGFbeta receptor/Smad pathway through gene ablation in mice. The International Journal of Developmental Biology. 2000;**44**(3):253-265

[20] Massague J, Chen YG. Controlling TGF-beta signaling. Genes & Development. 2000;**14**(6):627-644

[21] Badlani N, Oshima Y, Healey R, Coutts R, Amiel D. Use of bone morphogenic protein-7 as a treatment for osteoarthritis. Clinical Orthopaedics and Related Research. 2009;**467**(12):3221-3229

[22] Elshaier AM, Hakimiyan AA, Rappoport L, Rueger DC, Chubinskaya S. Effect of interleukin-1beta on osteogenic protein 1-induced signaling in adult human articular chondrocytes. Arthritis and Rheumatism. 2009;**60**(1):143-154

[23] Chubinskaya S, Hurtig M, Rueger DC. OP-1/BMP-7 in cartilage repair. International Orthopaedics. 2007;**31**(6):773-781

[24] Chubinskaya S, Hakimiyan A, Pacione C, Yanke A, Rappoport L, Aigner T, et al. Synergistic effect of IGF-1 and OP-1 on matrix formation by normal and OA chondrocytes cultured in alginate beads. Osteoarthritis and Cartilage. 2007;**15**(4):421-430

[25] Shen B, Wei A, Whittaker S, Williams LA, Tao H, Ma DD, et al. The role of BMP-7 in chondrogenic and osteogenic differentiation of human bone marrow multipotent mesenchymal stromal cells in vitro. Journal of Cellular Biochemistry. 2010;**109**(2):406-416

[26] Chubinskaya S, Otten L, Soeder S, Borgia JA, Aigner T, Rueger DC, et al. Regulation of chondrocyte gene expression by osteogenic protein-1. Arthritis Research & Therapy. 2011;**13**(2):R55

[27] Lires-Dean M, Carames B, Cillero-Pastor B, Galdo F, Lopez-Armada MJ, Blanco FJ. Anti-apoptotic effect of transforming growth factor-beta1 on human articular chondrocytes: Role of protein phosphatase 2A. Osteoarthritis and Cartilage. 2008;**16**(11):1370-1378

[28] Abe T, Yamada H, Nakajima H, Kikuchi T, Takaishi H, Tadakuma T, et al. Repair of full-thickness cartilage defects using liposomal transforming growth factor-beta1. Journal of Orthopaedic Science. 2003;**8**(1):92-101

[29] Diao H, Wang J, Shen C, Xia S, Guo T, Dong L, et al. Improved cartilage

regeneration utilizing mesenchymal stem cells in TGF-beta1 gene-activated scaffolds. Tissue Engineering. Part A. 2009;**15**(9):2687-2698

[30] Vincent TL, Hermansson MA, Hansen UN, Amis AA, Saklatvala J. Basic fibroblast growth factor mediates transduction of mechanical signals when articular cartilage is loaded. Arthritis and Rheumatism. 2004;**50**(2):526-533

[31] Im HJ, Li X, Muddasani P, Kim GH, Davis F, Rangan J, et al. Basic fibroblast growth factor accelerates matrix degradation via a neuro-endocrine pathway in human adult articular chondrocytes. Journal of Cellular Physiology. 2008;**215**(2):452-463

[32] Ellman MB, Yan D, Ahmadinia K, Chen D, An HS, Im HJ. Fibroblast growth factor control of cartilage homeostasis. Journal of Cellular Biochemistry. 2013;**114**(4):735-742

[33] Liu Z, Xu J, Colvin JS, Ornitz DM. Coordination of chondrogenesis and osteogenesis by fibroblast growth factor 18. Genes & Development. 2002;**16**(7):859-869

[34] Liu Z, Lavine KJ, Hung IH, Ornitz DM. FGF18 is required for early chondrocyte proliferation, hypertrophy and vascular invasion of the growth plate. Developmental Biology. 2007;**302**(1):80-91

[35] Nakao K, Kubota S, Doi H, Eguchi T, Oka M, Fujisawa T, et al. Collaborative action of M-CSF and CTGF/CCN2 in articular chondrocytes: Possible regenerative roles in articular cartilage metabolism. Bone. 2005;**36**(5):884-892

[36] Hoshijima M, Hattori T, Aoyama E, Nishida T, Yamashiro T, Takigawa M. Roles of heterotypic CCN2/CTGF-CCN3/NOV and homotypic CCN2-CCN2 interactions in expression of the differentiated phenotype of chondrocytes. The FEBS Journal. 2012;**279**(19):3584-3597

[37] Zhu S, Zhang B, Man C, Ma Y, Liu X, Hu J. Combined effects of connective tissue growth factor-modified bone marrow-derived mesenchymal stem cells and NaOH-treated PLGA scaffolds on the repair of articular cartilage defect in rabbits. Cell Transplantation. 2014;**23**(6):715-727

[38] Mullen LM, Best SM, Ghose S, Wardale J, Rushton N, Cameron RE. Bioactive IGF-1 release from collagen-GAG scaffold to enhance cartilage repair in vitro. Journal of Materials Science. Materials in Medicine. 2015;**26**(1):5325

[39] Tahimic CG, Wang Y, Bikle DD. Anabolic effects of IGF-1 signaling on the skeleton. Frontiers in Endocrinology (Lausanne). 2013;**4**:6

[40] Matsumoto T, Gargosky SE, Iwasaki K, Rosenfeld RG. Identification and characterization of insulin-like growth factors (IGFs), IGF-binding proteins (IGFBPs), and IGFBP proteases in human synovial fluid. The Journal of Clinical Endocrinology and Metabolism. 1996;**81**(1):150-155

[41] Uchimura T, Foote AT, Smith EL, Matzkin EG, Zeng L. Insulin-like growth factor II (IGF-II) inhibits IL-1beta-induced cartilage matrix loss and promotes cartilage integrity in experimental osteoarthritis. Journal of Cellular Biochemistry. 2015;**116**(12):2858-2869

[42] Hill TP, Spater D, Taketo MM, Birchmeier W, Hartmann C. Canonical Wnt/beta-catenin signaling prevents osteoblasts from differentiating into chondrocytes. Developmental Cell. 2005;**8**(5):727-738

[43] Reinhold MI, Kapadia RM, Liao Z, Naski MC. The Wnt-inducible transcription factor

Twist1 inhibits chondrogenesis. The Journal of Biological Chemistry. 2006;**281**(3):1381-1388

[44] Day TF, Guo X, Garrett-Beal L, Yang Y. Wnt/beta-catenin signaling in mesenchymal progenitors controls osteoblast and chondrocyte differentiation during vertebrate skeletogenesis. Developmental Cell. 2005;**8**(5):739-750

[45] Guo X, Mak KK, Taketo MM, Yang Y. The Wnt/beta-catenin pathway interacts differentially with PTHrP signaling to control chondrocyte hypertrophy and final maturation. PLoS One. 2009;**4**(6):ew6067

[46] Sato S, Kimura A, Ozdemir J, Asou Y, Miyazaki M, Jinno T, et al. The distinct role of the Runx proteins in chondrocyte differentiation and intervertebral disc degeneration: Findings in murine models and in human disease. Arthritis and Rheumatism. 2008;**58**(9):2764-2775

[47] Stricker S, Fundele R, Vortkamp A, Mundlos S. Role of Runx genes in chondrocyte differentiation. Developmental Biology. 2002;**245**(1):95-108

[48] Wang WJ, Sun C, Liu Z, Sun X, Zhu F, Zhu ZZ, et al. Transcription factor Runx2 in the low bone mineral density of girls with adolescent idiopathic scoliosis. Orthopaedic Surgery. 2014;**6**(1):8-14

[49] Duval E, Bouyoucef M, Leclercq S, Bauge C, Boumediene K. Hypoxia inducible factor 1 alpha down-regulates type i collagen through Sp3 transcription factor in human chondrocytes. IUBMB Life. 2016;**68**(9):756-763

[50] Fernandez MP, Young MF, Sobel ME. Methylation of type II and type I collagen genes in differentiated and dedifferentiated chondrocytes. The Journal of Biological Chemistry. 1985;**260**(4):2374-2378

[51] Haq SH. 5-Aza-2′-deoxycytidine acts as a modulator of chondrocyte hypertrophy and maturation in chick caudal region chondrocytes in culture. Anatomy & Cell Biology. 2016;**49**(2):107-115

[52] Zimmermann P, Boeuf S, Dickhut A, Boehmer S, Olek S, Richter W. Correlation of COL10A1 induction during chondrogenesis of mesenchymal stem cells with demethylation of two CpG sites in the COL10A1 promoter. Arthritis and Rheumatism. 2008;**58**(9):2743-2753

[53] Kumar D, Lassar AB. Fibroblast growth factor maintains chondrogenic potential of limb bud mesenchymal cells by modulating DNMT3A recruitment. Cell Reports. 2014;**8**(5):1419-1431

[54] Shen J, Abu-Amer Y, O'Keefe RJ, McAlinden A. Inflammation and epigenetic regulation in osteoarthritis. Connective Tissue Research. 2017;**58**(1):49-63

[55] Chabane N, Zayed N, Afif H, Mfuna-Endam L, Benderdour M, Boileau C, et al. Histone deacetylase inhibitors suppress interleukin-1beta-induced nitric oxide and prostaglandin E2 production in human chondrocytes. Osteoarthritis and Cartilage. 2008;**16**(10):1267-1274

[56] Young DA, Lakey RL, Pennington CJ, Jones D, Kevorkian L, Edwards DR, et al. Histone deacetylase inhibitors modulate metalloproteinase gene expression in chondrocytes and block cartilage resorption. Arthritis Research & Therapy. 2005;**7**(3):R503-R512

[57] Wienholds E, Kloosterman WP, Miska E, Alvarez-Saavedra E, Berezikov E, de Bruijn E, et al. MicroRNA expression in zebrafish

embryonic development. Science. 2005;**309**(5732):310-311

[58] Miyaki S, Nakasa T, Otsuki S, Grogan SP, Higashiyama R, Inoue A, et al. MicroRNA-140 is expressed in differentiated human articular chondrocytes and modulates interleukin-1 responses. Arthritis and Rheumatism. 2009;**60**(9):2723-2730

[59] Nakamura Y, He X, Kato H, Wakitani S, Kobayashi T, Watanabe S, et al. Sox9 is upstream of microRNA-140 in cartilage. Applied Biochemistry and Biotechnology. 2012;**166**(1):64-71

[60] Florine EM, Miller RE, Liebesny PH, Mroszczyk KA, Lee RT, Patwari P, et al. Delivering heparin-binding insulin-like growth factor 1 with self-assembling peptide hydrogels. Tissue Engineering. Part A. 2015;**21**(3-4):637-646

[61] Florine EM, Miller RE, Porter RM, Evans CH, Kurz B, Grodzinsky AJ. Effects of dexamethasone on mesenchymal stromal cell chondrogenesis and aggrecanase activity: Comparison of agarose and self-assembling peptide scaffolds. Cartilage. 2013;**4**(1):63-74

[62] Roach BL, Kelmendi-Doko A, Balutis EC, Marra KG, Ateshian GA, Hung CT. Dexamethasone release from within engineered cartilage as a chondroprotective strategy against interleukin-1alpha. Tissue Engineering. Part A. 2016;**22**(7-8):621-632

Chapter 3

Nonsurgical Strategies for the Treatment of Temporomandibular Joint Disorders

Juan L. Cobo, Manuela Cabrera-Freitag, Teresa Cobo, Juan D. Muriel, Luis M. Junquera, Juan Cobo and José A. Vega

Abstract

Temporomandibular disorders are common maxillofacial disturbs of different etiologies (traumatic, inflammatory, degenerative, or congenital) that course with pain and dysfunctions of the temporomandibular joint. The treatment of these disorders includes systematically administered drugs (especially nonsteroid anti-inflammatory drugs and corticoids), physical therapies, and minimally invasive therapies that require intraarticular injections. These techniques are directed to clean or drain the articular cavity, to deliver intraarticularly drugs, biologically active compounds (as platelet-rich plasma), or to enhance lubrication (hyaluronic acid). Moreover, minimally invasive strategies are used in regenerative medicine for to deliver cells and stem cells, and nano- or micro-biomaterials. Surgery of temporomandibular disorders is only used in grave diseases that require arthrodesis or remotion of the temporomandibular joint. This review updates the nonsurgical therapeutic strategies to treat temporomandibular disorders, focusing the attention in the articular delivery or hyaluronic acid and platelet-rich plasma, two minimally invasive widely used at present.

Keywords: temporomandibular disorders, minimally invasive therapies, hyaluronic acid, platelet-rich plasma, regenerative medicine

1. Introduction: temporomandibular joint disorders

The temporomandibular joint (TMJ) is the only dynamic articulation of the head and present unique anatomical, structural, and biochemical characteristics. Up to 40–50% of the population suffers different pathologies of TMJ [1, 2] that requires therapeutic interventions by different medical and paramedical specialists and represents an increasing social and psychosocial impairment [3].

TMJ disorders (TMD) are a class of degenerative musculoskeletal conditions associated with morphological and functional deformities, which clinically result in pain and TMJ dysfunctions (impairment in mastication, speech, and facial expression) (see for a review [4]). Moreover, when TMD affect young subjects during growth, it can cause asymmetry of the facial skeleton [5]. In agreement with the above definition, TMD comprise a heterogeneous group of pathologies involving the TMJ, the associated jaw muscles, or both [6]. Up to 40–50% of the population suffers TMD [2], and up to 70% of them suffer TMD directly related to the articular disc [1].

IntechOpen

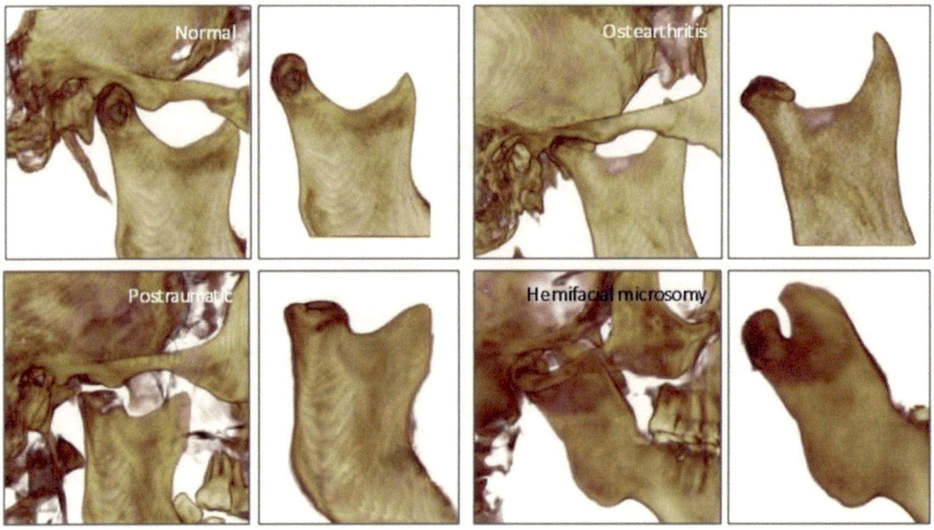

Figure 1.
Cone bean CT of the right adult TMJ in normal conditions, osteoarthritis, posttraumatic, and hemifacial microsomy. Images obtained from Instituto Asturiano de Odontolgía, Oviedo, Spain.

The etiology of TMD can be traumatic, inflammatory, and congenital [6]. However, the primary TMD are degenerative inflammatory or noninflammatory diseases, that is, osteoarthritis or arthrosis, respectively [6]. Typical osteoarthritic changes include alterations in shape and size of TMJ components (flattened fossa, reduced articular eminence, decreased condylar volume, and thickened disc), abrasion of articular cartilage, and thickening and remodeling of the subchondral bone that leads to morphological deformity and dysfunction (**Figure 1**) [4].

2. Brief summary of the anatomy and structure of the temporomandibular joint

TMJ is a bilateral diarthrodial joint formed by the condylar head of the mandible and the glenoid fossa (or mandibular fossa) of the temporal bone, surrounded by a fibrous capsule reinforced laterally (lateral temporomandibular ligament) and two extracapsular ligaments (sphenomandibular and stylomandibular). Interposed between the mandibular condyle and the temporal bone, there is an articular disc of fibrocartilage attached partially to the bones and the capsule that incompletely divides the TMJ into two chambers: upper or temporodisc chamber, and lower chamber or disc-condylar chamber [7].

One differential characteristic of TMJ is that the cartilage covering the articular surfaces is not hyaline cartilage, as in other diarthrosis, but a fibrocartilaginous tissue [8]. It can be regarded as a modified fibrous periosteum with an underlying proliferative zone that differentiates into fibrocartilage [9]. In TMJ articular cartilage, from the surface to the bone, two different zones are considered: the *fibrous zone* and the *fibrocartilage zone,* which can be subdivided into *proliferative* and *hypertrophic zones*. The fibrous zone contains fibroblasts, and the extracellular matrix (ECM) consists of type I collagen, type II collagen at residual levels, and versican-like chondroitin sulfate-based proteoglycan. The cells of the fibrocartilage zone are fibroblasts and chondrocytes, and the ECM is rich in type II collagen, but also contains type I and type X collagen, and aggrecan (**Figure 2** [10]).

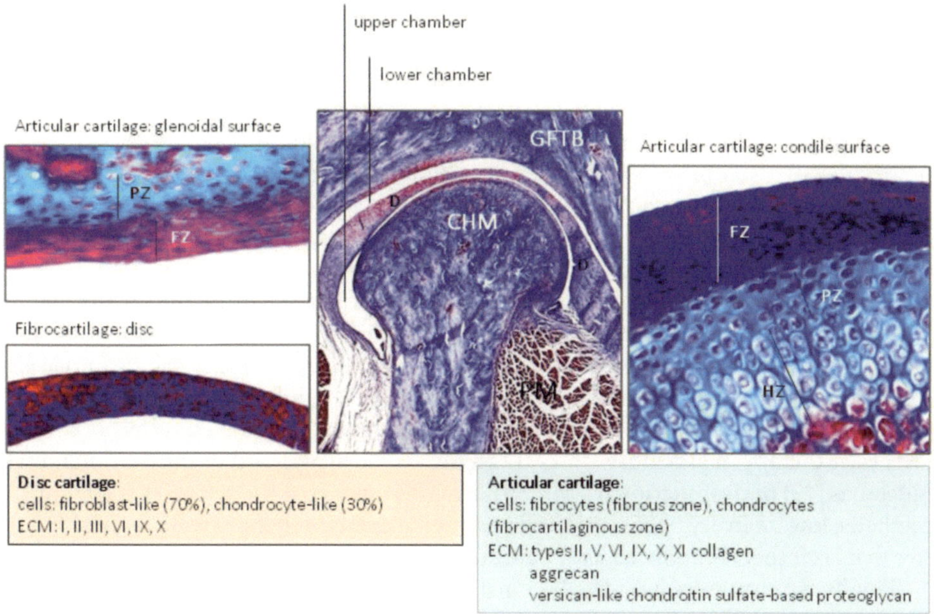

Figure 2.
Organization of the rat temporomandibular joint. CHM: condylar head of the mandible and GFTB: glenoidal fossa of temporal bone. FZ: fibrous zone, HZ: hypertrophic zone, and PZ: proliferative zone. The boxes contain the cells and the main biochemical characteristics of the articular cartilage and the articular disc.

The fibrocartilage forming the articular disc consists of several populations of cells: fibroblast-like and chondrocyte-like cells, 70 and 30%, respectively [11]. In ECM, type I collagen predominates but other collagens (types II, III, VI, IX, and XII) are present [12, 13], and also contains glycosaminoglycans (**Figure 2**) [14].

Along the articular temporal surface, each mandibular condyle has a wide motion range, consisting of both rotation and translation. TMJ movements are involved in facial expressions, talking, drinking, and eating [15, 16].

3. Treatment of TMD

The treatment of TMD varies according to the etiology and severity of the lesion and can be divided into noninvasive, minimally invasive, and invasive, all of them focused to alleviate the symptoms, and repair or replace the pathologic TMJ structures.

Invasive treatments that are always surgical are out of the scope of this chapter, and represent the unique option for patients suffering severe TMD like traumatisms, neoplasia, or developmental malformations. In most cases, it is necessary to perform an arthrotomy to restoring joint tissues or replace TMJ with autogenous or alloplastic material. In the TMD due to disc alterations, surgical repositioning, the removal (discectomy [17]), or replacement [14, 18] have been used with variable efficacy.

The noninvasive treatments include drugs, occlusal orthodontics, physical therapy, or acupuncture. The used drugs are analgesics, NSAIDs, anxiolytics, muscle relaxants, and opioids, all administered systematically [19–21]. The occlusal orthodontics and occlusal splint are widely used for the treatment of TMD, but their effectiveness remains questionable. At present, there is no evidence for a cause-effect relationship between orthodontic treatment and TMD, or that such treatment might improve or prevent them [22]. Furthermore, there is insufficient evidence either for or against the use of stabilization splint therapy for the treatment of the pain of TMD [23]. The same applies for the oral appliances that might reduce pain and assist in maintaining stable function between jaw posture, muscle function, and temporomandibular joint stability [24] although TMD can result as a side effect from use those devices [25].

The physical therapies for TMD include different techniques like exercises, neuromuscular stabilization, electrotherapy and transcutaneous electrical nerve stimulation (TENS), low-intensity ultrasound, and low-level laser therapy. These methods are easily applicable and have demonstrated efficiency in some cases of TMD especially those of muscular origin.

Physiotherapy is commonly employed in the treatment of TMDs, but its relative efficacy is unclear, and most methods (short-wave diathermy, megapulse, ultrasound, and soft laser) have similar beneficial effects (range 70.4–77.7%) [25, 26]. In any case, a mixed approach of therapies has impact on reducing pain, increasing range of motion, but lacks a significant impact for functional improvement [27, 28]. The effect of low-level laser therapy in patients with TMD seems to relieve pain and improves functional outcomes [29] or dysfunctional TMJ [30]. And in comparing the effects of different methods, low intensity ultrasound and traditional exercise therapy were more effective that laser therapy reduced TMJ pain and trismus after oncologic surgery [31].

Finally, acupuncture has also demonstrated to reduce symptoms associated with TMD. Meta-analysis noted moderate evidence that acupuncture is effective to reduce symptoms associated with TMD, and trials with adequate sample sizes are necessary that address the long-term efficacy or effectiveness of acupuncture [32, 33].

As a whole, and despite limited evidence, physical therapy can be an effective treatment option for TMD, with jaw exercise (79%), ultrasound (52%), manual therapy (MT) (48%), acupuncture (41%), and laser therapy (15%) as the most effective modalities for managing TMD [34].

The minimally invasive treatments include the therapies that require intraarticular injections, arthrocentesis, or arthroscopy. They are used to clean or drain the articular cavity, to deliver intraarticularly active substance like drugs (NSAIDs and corticosteroids [35–37], biologically-active compounds (for example platelet-rich plasma [38]), or enhance lubrication (hyaluronic acid (**Figure 3**) [35]). Current clinical therapies using intraarticular injections are effective in pain relief at an early stage of disease but fail to alleviate chronic pain.

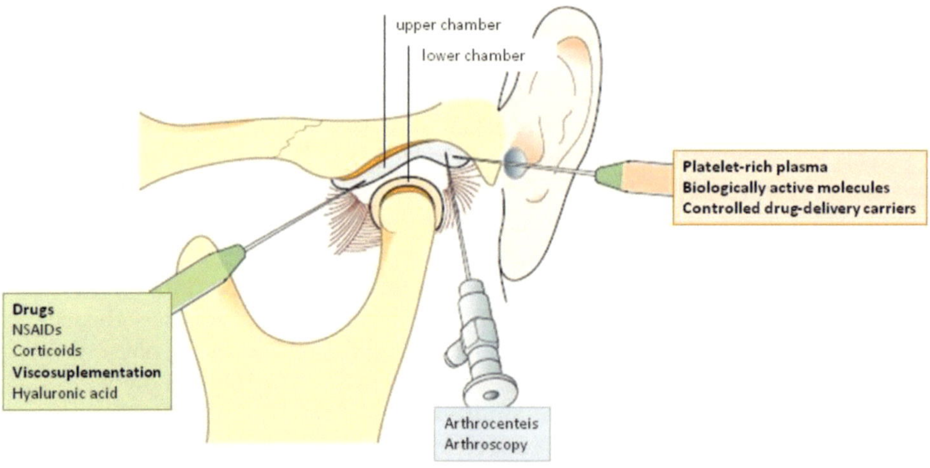

Figure 3.
Schematic representation of the minimally invasive methods and the compounds delivered in TMJ intraarticularly. Modified of https://pocketdentistry.com/33-temporomandibular-joint-surgery-including-arthroscopy/.

Furthermore, minimally invasive strategies are now used in regenerative medicine for treatment of TMD, to deliver cells and stem cells, nano- or micro-biomaterials, carriers of drugs with controlled release [39–41]. Actually, it is also of interest the delivery of therapeutic molecules through the use of nanoparticles- (NP-BDS) and microparticles- (MP-BDS) based delivery system that can release therapeutic molecules in a controlled or sustained manner and target specific cells (chondrocytes and synoviocytes). The nano- and microparticles interact with cells at the intra- and extracellular space depending on their size.

NP-BDS are solid or colloidal particles with sizes ranging from tens to hundreds of nanometers, which are endocyted and enter into the cytoplasm cells where they release small-sized biomolecules intracellularly [40, 41].

MP-BDS are synthetic or natural polymers spherically shaped with sizes ranging from ten to hundreds of micrometers and are suitable to deliver large drugs or biomolecules acting on the cell surface, thus extracellularly; they serve as vehicles for corticoids and NSAIDs. In addition, microparticles can also release biomolecules and deliver stem cells (see [41]).

4. Intraarticular delivery of hyaluronic acid

Hyaluronic acid or hyaluronan (HA) is a component of ECM and the body fluids, including the synovial fluid that organizes proteoglycans and other proteins on the cell membrane surface through noncovalent unions; in the fluids, it is responsible for their rheological properties. Structurally, HA is a glycosaminoglycan polymer formed by repeated sequence of D-glycuronic acid and N-acetyl-D-glycosamine linked by means of alternant β-1, β-1, 4, and 3 glycoside links. HA plays a key role in the physiology of diarthrosis especially in the articular cartilage as well as in the maintenance of synovial fluid viscosity, thus in viscoelasticity and lubrication. It is synthesized by the synoviocytes and has a molecular weight of about 6000–7000 kDa.

Most of the inflammatory and degenerative joint diseases course with increased local concentrations of pro-inflammatory molecules and proteases that degrade HA originating from small HA-fragments with a low-molecular weight. Consequently, in those diseases, there is a reduction in the viscosity and lubrication properties of the synovial fluid and a dramatic change in the biological receptor-mediated effects of HA. Moreover, the resulting small fragments acting through different membrane receptors can stimulate the inflammatory responses in the synovial membrane and the lesions in the articular cartilage [42, 43]. Therefore, one of the therapeutic strategies for the treatment of some joint diseases is to restore the rheological properties of the synovial fluid [44] and the joint homeostasis [45] throughout the intraarticular delivery of HA.

HA plays a key role in the pathogenesis of the degenerative and traumatic joint diseases acting as a pro-inflammatory or anti-inflammatory molecule, stimulating or inhibiting cellular migration, division, and differentiation [46]. The final effects depend both on the state of the tissue (expression of HA receptors, phase of the cell cycle, and signaling pathways) [47] and the characteristics of the HA (tridimensional structure and the size of the HA molecule) [48–50].

The intraarticular administration of exogenous HA is called "viscosupplementation," and it is focused to restore the rheological properties of the synovial fluid and to block the generative processes. Until now, the effectiveness of intraarticular administration of HA offers discordant results [51, 52]. Nevertheless, the meta-analysis of treatments that used intraarticular HA and the European Society for Clinical and Economic Aspects of Osteoarthritis recommends the use of intraarticular injections of HA in absence of response to conventional anti-inflammatory drugs, since it improves the functionality of the joint and diminishes pain [53, 54].

The beneficial effects of viscosupplementation with HA in TMD have not been probed satisfactory and are not more effective that of corticosteroids and NSAIDs [35, 55, 56]. Also, although there was no significant difference between the effectiveness HA and corticoids intraarticular injections, there was some evidence that HA was better than placebo [57]. However, most studies report a decrease in pain levels independently by the TMD [58]. On the other hand, it seems that HA regulates various inflammatory mediators in osteoarthritis in the TMJ [59]. In any case, at present, there is insufficient, consistent evidence to either support or refute the use of HA for treating patients with TMD.

5. Intraarticular delivery of platelet-rich plasma

Platelet-rich plasma (PRP; blood plasma that has been enriched with platelets) therapies have emerged as a potential approach to enhance tissue repair and regeneration, and have demonstrated to be a safe, resourceful, and effective treatment. They are based on the delivery of growth factors and cytokines from anuclear platelets that can stimulate the healing of various tissues as a consequence of activation of migratory and local cells [60, 61]. Nevertheless, because PRP is autologous, the concentration of the PRP components differs according to the physiological conditions and clinical diseases of patients [62].

The biological effects of PRP are largely attributed to the platelet secretome and some plasma signaling proteins. In fact, the α-granules of platelets within PRP release numerous growth factors and cytokines (TGF-α, TGF-β, HGF, IL-6, EGF, FGF-2, IGF-1, VEGF, and interleukin β1). Moreover, PRP contains proteases, biologically active amines, and cell adhesion molecules such as fibrin, fibronectin, and vitronectin [60]. All those molecules are involved in repair and regeneration processes, including anti-apoptosis, cell proliferation, differentiation, migration, angiogenesis, and the synthesis of ECM in both normal and pathological conditions [63]. Cells within the joint add to this milieu by secreting additional biologically active molecules in response to PRP.

PRP is currently used in patients with chronic joint pain caused by progressive cartilage degeneration of the synovial joints. The anti-inflammatory effects are carried out through its effects on nuclear factor κB signaling pathway (including synoviocytes, macrophages, and chondrocytes), but also by reducing TNF-α and IL-1β [64]. A systematic review and meta-analysis related to the clinical efficacy of intraarticular PRP injection in patients with osteoarthritis have shown significant clinical improvements [65, 66].

Recently, Kütük et al. [67] and Hegab et al. [68] reported that an intraarticular PRP injection is an effective treatment for TMJ osteoarthritis through the regeneration of fibrocartilage and cartilage, bone repair in the TMJ. Moreover, PRP has long-term analgesic effects in most patients with painful TMJ [69, 70]. Nevertheless, a randomized clinical trial in patients with TMJ osteoarthritis suggests that arthrocentesis plus PRP injections is not superior to arthrocentesis alone or combined with HA injection, and PRP does not add any significant improvement to clinical outcomes after surgery in patients with advanced internal derangement of the TMJ [71, 72]. Thus, PRP injection should not be considered as a first-line treatment for TMD, and arthrocentesis plus HA injection would appear to be more acceptable [73]. Nevertheless, other authors observed that PRP performed well than HA in the treatment of TMJ osteoarthritis in terms of pain reduction for the treatment of reducible disc displacement of the TMJ [68, 72, 74]. Future studies will focus on the synergistic actions of HA and PRP in the treatment of TMJ osteoarthritis as in other joints.

6. Tissue engineering

In recent past years, detailed and exhaustive reviews have been published covering all the relevant data about the experimental [75], technical aspects, and indications of tissue engineering in TMJ [76–81]. Therefore, this section only summarizes the most relevant aspects of tissue engineering of TMJ using minimally invasive techniques. In the last two decades, new studies have contributed to understand what are the appropriate scaffolds, cells and biological for TMJ diseases, and all these advancements are based on the perfectly known structures of the different joint constituents.

Traditionally, the principal elements of tissue engineering-based regenerative strategies are scaffolds, cells, and biological stimuli. Those used in TMJ are summarized in **Table 1**. Although through invasive methods all strategies are possible to regenerate TMJ components when minimally invasive techniques are used, two methods are possible in cartilage and bone engineering: *in situ* tissue engineering incorporating an acellular scaffold matrix that attract and fix local cells thus guiding the process of regeneration and *ex vivo* cell seeding on the scaffold that initiates and regulates the regenerative mechanisms [101]. On the other hand, to induce more rapid ECM synthesis, scaffolds can be embedded with growth factors. Also,

Tissue		References
Condylar cartilage		
Scaffolds	Hyaluronic acid hydrogels	[82]
	Agarose	[83]
	Poly-vinyl alcohol	[84]
	Poly-l-lactic-coglycolic acid	[85]
Cells	Chondrocytes	[86, 87]
	Synovial stem cells	[88, 89]
	Bone marrow mesenchymal stem cells	[88, 90]
	Adipose stem cells	[91]
	Tooth-derived stem cells	[92]
Articular disc		
Scaffolds	Polyglycerol sebacate	[93]
	Poly-glycolic acid	[94, 95]
	Poly-l-lactic acid	[77, 96]
	Poly(glycerol sebacate)	[93]
	Polycaprolactone	[97]
	Polytetrafluorethylene monofilaments + poly-l-lactic acid monofilaments + polyamide monofilaments + natural bone	[98]
	Chitosan	[99]
	Alginate hydrogels	[94]
	Decellularized ECM	[100]
Cells	Dermal fibroblasts	[95]
	Synovial stem cells	[88]

Table 1.
Scaffolds and cells used in TMJ tissue engineering.

intraarticular injection of cells or local delivery of biologically active molecules can be a strategy, but these cannot be regarded properly as tissue engineering.

Scaffolds serve as a supportive structure to the engineered tissues. As a rule, the used scaffolds must promote the differentiation of cells into chondrocytes and stimulate the synthesis of cartilaginous ECM. Both natural and synthetic scaffolds have been experimented for engineering the TMJ (**Table 1**). Nevertheless, the most suitable approach should be reconstructed for both full articulating surfaces by stabilizing scaffolds on the articular surfaces to be regenerated and autologous chondrocytes within the scaffold. But in the case of TMJ, the reconstruction of the disc is also important. Nevertheless, as the replacement of the articular disc does not seem to be feasible at the current state of tissue engineering, lining the articular fossa with resistant engineered cartilage tissue would be an alternative in patients after discectomy [78].

Diverse cells have been used in TMJ tissue engineering (see **Table 1**) within different scaffolds. The local delivery of mesenchymal stem cells (MSCs) within TMJ has proved to have beneficial effects on TMJ degenerative diseases [79]. Furthermore, another strategy would be stimulating the resident mesenchymal stem cells present in the synovial layer [102] and synovial fluid of TMJ [103]. MSCs are able to secrete bioactive molecules, such as growth factors, cytokines, and chemokines, which exert their biological role under injury conditions [104].

Growth factors help tissue regeneration promoting the differentiation and proliferation of cells and supporting ECM synthesis and specialization. Thus, the incorporation of growth factors to the scaffolds, the direct intraarticular delivery of growth factors, or stimulating the exogenous or resident cells to secrete and release growth factor can result in an improvement of tissue regeneration. Various technologies for incorporation of growth factors into scaffolds are possible. At present, the three key growth factors for TMJ regeneration are bFGF, IGF-1, and TGF-β1 [105]. However, fibrochondrocytes from mandibular condyle are less responsive to IGF-1 than hyaline chondrocytes [86]. TGF-β1 stimlates cell proliferation, and on the production of ECM in TMJ disc implants [106], and TGF-β1 and IGF-1 acting together promote cellular proliferation and secretion of type I collagen and glycosaminoglycans [107]. In culture, bFGF increased the proliferation of fibrochondrocytes from mandibular condyle more than TGF-β1 and IGF-1 [108]. Finally, PDGF significantly increases the proliferation rate of the TMJ-disc-derived cells, collagen, and hyaluronic acid synthesis in engineered TMJ disc [109].

Another source of bioactive molecules to be delivered into TMJ is the MSC-conditioned medium collectively known as the MSC secretome. It contains trophic factors and various MSC-based clinical trials that have revealed that transplanted MSCs exert their biological functions through trophic modulations rather than differentiation potential [110]. Similar properties have the secretome of the periodontal ligament-derived MSCs [111]. Finally, exosomes, cell-secreted nano-sized vesicles covered by the bilipid membrane, containing a myriad of regulatory components including microRNAs (miRNAs), mRNAs, and proteins [112], could be in the future a reliable possibility to stimulate TMJ regeneration.

Author details

Juan L. Cobo[1,2], Manuela Cabrera-Freitag[1], Teresa Cobo[3,4], Juan D. Muriel[3], Luis M. Junquera[2,3], Juan Cobo[3,4] and José A. Vega[1,5]*

1 Departamento de Morfología y Biología Celular, Universidad de Oviedo, Spain

2 Servicio de Cirugía Máxilofacial, Hospital Universitario Central de Asturias, Oviedo, Spain

3 Departamento de Cirugía y Especialidades Médico-Quirúrgicas, Universidad de Oviedo, Spain

4 Instituto Asturiano de Odontología, Oviedo, Spain

5 Facultad de Ciencias de la Salud, Universidad Autónoma de Chile, Santiago de Chile, Chile

*Address all correspondence to: javega@uniovi.es

References

[1] Manfredini D, Guarda-Nardini L, Winocur E, Piccotti F, Ahlberg J, Lobbezoo F. Research diagnostic criteria for temporomandibular disorders: A systematic review of axis I epidemiologic findings. Oral Surgery, Oral Medicine, Oral Pathology, Oral Radiology, and Endodontics. 2011;**112**:453-462

[2] Gopal SK, Shankar R, Vardhan BH. Prevalence of temporo-mandibular disorders in symptomatic and asymptomatic patients: A cross-sectional study. International Journal of Advanced Health Sciences. 2014;**1**:14-20

[3] De La Torre Canales G, Câmara-Souza MB, Muñoz Lora VRM, Guarda-Nardini L, Conti PCR, Rodrigues Garcia RM, et al. Prevalence of psychosocial impairment in temporomandibular disorder patients: A systematic review. Journal of Oral Rehabilitation. 2018;**45**:881-889

[4] Murphy MK, MacBarb RF, Wong ME, Athanasiou KA. Temporomandibular joint disorders: A review of etiology, clinical management, and tissue engineering strategies. The International Journal of Oral & Maxillofacial Implants. 2013;**28**:e393

[5] Roberts WE, Stocum DL. Part II: Temporomandibular joint (TMJ)-regeneration, degeneration, and adaptation. Current Osteoporosis Reports. 2018;**16**:369-379

[6] Tanaka E, Detamore MS, Mercuri LG. Degenerative disorders of the temporomandibular joint: Etiology, diagnosis, and treatment. Journal of Dental Research. 2008;**87**:296-307

[7] Alomar X, Medrano J, Cabratosa J, Clavero JA, Lorente M, Serra I, et al. Anatomy of the temporomandibular joint. Seminars in Ultrasound, CT, and MR. 2007;**28**:170-183

[8] Robinson PD. Articular cartilage of the temporomandibular joint: Can it regenerate? Annals of the Royal College of Surgeons of England. 1993;**75**:231-236

[9] Stocum DL, Roberts WE. Part I: Development and physiology of the temporomandibular joint. Current Osteoporosis Reports. 2018;**16**:360-368

[10] Kuroda S, Tanimoto K, Izawa T, Fujihara S, Koolstra JH, Tanaka E. Biomechanical and biochemical characteristics of the mandibular condylar cartilage. Osteoarthritis and Cartilage. 2009;**17**:1408-1415

[11] Detamore MS, Hegde JN, Wagle RR, Almarza AJ, Montufar-Solis D, Duke PJ, et al. Cell type and distribution in the porcine temporomandibular joint disc. Journal of Oral and Maxillofacial Surgery. 2006;**64**:243-248

[12] Minarelli AM, Del Santo Júnior M, Liberti EA. The structure of the human temporomandibular joint disc: A scanning electron microscopy study. Journal of Orofacial Pain. 1997;**11**:95-100

[13] Kalpakci KN, Willard VP, Wong ME, Athanasiou KA. An interspecies comparison of the temporomandibular joint disc. Journal of Dental Research. 2011;**90**:193-198

[14] Willard VP, Zhang L, Athanasiou KA. Tissue engineering of the temporomandibular joint. In: Ducheine P, editor. Comprehensive Biomaterials: Tissue and Organ Engineering. Vol. 5. Elsevier Science. 2011. pp. 221-235

[15] Naeije M, Hofman N. Biomechanics of the human temporomandibular

joint during chewing. Journal of Dental Research. 2003;**82**:528-531

[16] Tanaka E, Koolstra J. Biomechanics of the temporomandibular joint. Journal of Dental Research. 2008;**87**:989-991

[17] Miloro M, Henriksen B. Discectomy as the primary surgical option for internal derangement of the temporomandibular joint. Journal of Oral and Maxillofacial Surgery. 2010;**68**:782-789

[18] Gerbino G, Zavattero E, Bosco G, Berrone S, Ramieri G. Temporomandibular joint reconstruction with stock and custom-made devices: Indications and results of a 14-year experience. Journal of Cranio-Maxillo-Facial Surgery. 2017;**45**:1710-1715

[19] Liu F, Steinkeler A. Epidemiology, diagnosis, and treatment of temporomandibular disorders. Dental Clinics of North America. 2013;**57**:465-479

[20] Heir GM. The efficacy of pharmacologic treatment of temporomandibular disorders. Oral and Maxillofacial Surgery Clinics of North America. 2018;**30**:279-285

[21] Ingawalé S, Goswami T. Temporomandibular joint: Disorders, treatments, and biomechanics. Annals of Biomedical Engineering. 2009;**37**:976-996

[22] Fernández-González FJ, Cañigral A, López-Caballo JL, Brizuela A, Moreno-Hay I, Del Río-Highsmith J, et al. Influence of orthodontic treatment on temporomandibular disorders. A systematic review. Journal of Clinical and Experimental Dentistry. 2015;7(2):e320-e327

[23] Al-Ani MZ, Davies SJ, Gray RJ, Sloan P, Glenny AM. Stabilisation splint therapy for temporomandibular pain dysfunction syndrome. Cochrane Database of Systematic Reviews. 2004;(1):CD002778. DOI: 10.1002/14651858.CD002778.pub2

[24] Greene CS, Menchel HF. The use of oral appliances in the management of temporomandibular disorders. Oral and Maxillofacial Surgery Clinics of North America. 2018;**30**:265-277

[25] Gray RJ, Quayle AA, Hall CA, Schofield MA. Physiotherapy in the treatment of temporomandibular joint disorders: A comparative study of four treatment methods. British Dental Journal. 1994;**176**:257-261

[26] McNeely ML, Armijo Olivo S, Magee DJ. A systematic review of the effectiveness of physical therapy interventions for temporomandibular disorders. Physical Therapy. 2006;**86**:710-725

[27] Armijo-Olivo S, Pitance L, Singh V, Neto F, Thie N, Michelotti A. Effectiveness of manual therapy and therapeutic exercise for temporomandibular disorders: Systematic review and meta-analysis. Physical Therapy. 2016;**96**:9-25

[28] Dickerson SM, Weaver JM, Boyson AN, Thacker JA, Junak AA, Ritzline PD, et al. The effectiveness of exercise therapy for temporomandibular dysfunction: A systematic review and meta-analysis. Clinical Rehabilitation. 2017;**31**:1039-1048

[29] Xu GZ, Jia J, Jin L, Li JH, Wang ZY, Cao DY. Low-level laser therapy for temporomandibular disorders: A systematic review with meta-analysis. Pain Research & Management. 2018;**2018**:4230583

[30] Alan H, Yolcu U, Koparal M, Ozgur C, Ozturk SA, Malkoc S. Evaluation of the effects of the low-level laser therapy

on swelling, pain, and trismus after removal of impacted lower third molar. Head & Face Medicine. 2016;**12**:25

[31] Elgohary HM, Eladl HM, Soliman AH, Soliman ES. Effects of ultrasound, laser and exercises on temporomandibular joint pain and trismus following head and neck cancer. Annals of Rehabilitation Medicine. 2018;**42**:846-853

[32] Cho SH, Whang WW. Acupuncture for temporomandibular disorders: A systematic review. Journal of Orofacial Pain. 2010;**24**:152-162

[33] Jung A, Shin BC, Lee MS, Sim H, Ernst E. Acupuncture for treating temporomandibular joint disorders: A systematic review and meta-analysis of randomized, sham-controlled trials. Journal of Dentistry. 2011;**39**:341-350

[34] Rashid A, Matthews NS, Cowgill H. Physiotherapy in the management of disorders of the temporomandibular joint; perceived effectiveness and access to services: A national United Kingdom survey. The British Journal of Oral & Maxillofacial Surgery. 2013;**51**:52-57

[35] Machado E, Bonotto D, Cunali PA. Intra-articular injections with corticosteroids and sodium hyaluronate for treating temporomandibular joint disorders: A systematic review. Dental Press Journal of Orthodontics. 2013;**18**:128-133

[36] Dym H, Bowler D, Zeidan J. Pharmacologic treatment for temporomandibular disorders. Dental Clinics of North America. 2016;**60**:367-379

[37] Gopalakrishnan V, Nagori SA, Roy Chowdhury SK, Saxena V. The use of intra-articular analgesics to improve outcomes after temporomandibular joint arthrocentesis: A review. Oral and Maxillofacial Surgery. 2018;**22**:357-364

[38] Zotti F, Albanese M, Rodella LF, Nocini PF. Platelet-rich plasma in treatment of temporomandibular joint dysfunctions: Narrative review. International Journal of Molecular Sciences. 2019;**20**:pii: E277

[39] Huang G, Zhang Z. Micro- and nano-carrier mediated intra-articular drug delivery systems for the treatment of osteoarthritis. Journal of Nanotechnology. 2012;**2012**:748909

[40] Mountziaris PM, Sing DC, Mikos AG, Kramer PR. Intra-articular microparticles for drug delivery to the TMJ. Journal of Dental Research. 2010;**89**:1039-1044

[41] Dashnyam K, Lee JH, Mandakhbayar N, Jin GZ, Lee HH, Kim HW. Intra-articular biomaterials-assisted delivery to treat temporomandibular joint disorders. Journal of Tissue Engineering. 2018;**9**:2041731418776514

[42] Campo GM, Avenoso A, D'Ascola A, Prestipino V, Scuruchi M, Nastasi G, et al. Inhibition of hyaluronan synthesis reduced inflammatory response in mouse synovial fibroblasts subjected to collagen-induced arthritis. Archives of Biochemistry and Biophysics. 2012;**518**:42-52

[43] Campo GM, Avenoso A, D'Ascola A, Scuruchi M, Nastasi G, Micali A, et al. The SOD mimic MnTM-2-PyP(5+) reduces hyaluronan degradation-induced inflammation in mouse articular chondrocytes stimulated with Fe (II) plus ascorbate. The International Journal of Biochemistry & Cell Biology. 2013;**45**:1610-1619

[44] Goldberg VM, Buckwalter JA. Hyaluronans in the treatment of osteoarthritis of the knee: Evidence for disease-modifying activity. Osteoarthritis and Cartilage. 2005;**13**:216-224

[45] Rani N, Sabbioni G, Mazzotta A, Rocchi M, Stagni C, Filanti M, et al. Infiltrative therapy as conservative treatment in hip osteoarthritis: A literature review. Hip International. 2016;**26**(Suppl 1):8-13

[46] Dicker KT, Gurski LA, Pradhan-Bhatt S, Witt RL, Farach-Carson MC, Jia X. Hyaluronan: A simple polysaccharide with diverse biological functions. Acta Biomaterialia. 2014;**10**:1558-1570

[47] Petrey AC, de la Motte CA. Hyaluronan, a crucial regulator of inflammation. Frontiers in Immunology. 2014;**5**:101

[48] Jiang D, Liang J, Noble PW. Hyaluronan as an immune regulator in human diseases. Physiological Reviews. 2011;**91**:221-264

[49] Stern R, Asari AA, Sugahara KN. Hyaluronan fragments: An information-rich system. European Journal of Cell Biology. 2006;**85**:699-715

[50] Maytin EV. Hyaluronan: More than just a wrinkle filler. Glycobiology. 2016;**26**:553-559

[51] Health Quality Ontario. Intra-articular viscosupplementation with hylan g-f 20 to treat osteoarthritis of the knee: An evidence-based analysis. Ontario Health Technology Assessment Series. 2005;**5**:1-66

[52] Newberry SJ, Fitzgerald JD, Maglione MA, O'Hanlon CE, Booth M, Motala A, et al. Systematic Review for Effectiveness of Hyaluronic Acid in the Treatment of Severe Degenerative Joint Disease (DJD) of the Knee. Rockville, MD: Agency for Healthcare Research and Quality (US); AHRQ Technology Assessments; 2015. Available from: https://www.cms.gov/medicare-coverage database/details/technology-assessments-details.aspx?TAId=101&bc=AAAQAAAAAAAAAA%3D%3D&

[53] Bannuru RR, Osani M, Vaysbrot EE, McAlindon TE. Comparative safety profile of hyaluronic acid products for knee osteoarthritis: A systematic review and network meta-analysis. Osteoarthritis and Cartilage. 2016;**24**:2022-2041

[54] Maheu E, Rannou F, Reginster JY. Efficacy and safety of hyaluronic acid in the management of osteoarthritis: Evidence from real-life setting trials and surveys. Seminars in Arthritis and Rheumatism. 2016;**45** (4 Suppl):S28-S33

[55] Goiato MC, da Silva EV, de Medeiros RA, Túrcio KH, Dos Santos DM. Are intra-articular injections of hyaluronic acid effective for the treatment of temporomandibular disorders? A systematic review. International Journal of Oral and Maxillofacial Surgery. 2016;**45**:1531-1537

[56] Ferreira N, Masterson D, Lopes de Lima R, de Souza Moura B, Oliveira AT, Kelly da Silva Fidalgo T, et al. Efficacy of viscosupplementation with hyaluronic acid in temporomandibular disorders: A systematic review. Journal of Cranio-Maxillo-Facial Surgery. 2018;**46**:1943-1952

[57] Moldez MA, Camones VR, Ramos GE, Padilla M, Enciso R. Effectiveness of intra-articular injections of sodium hyaluronate or corticosteroids for intracapsular temporomandibular disorders: A systematic review and meta-analysis. Journal of Oral & Facial Pain and Headache. 2018;**32**:53-66

[58] Manfredini D, Piccotti F, Guarda-Nardini L. Hyaluronic acid in the treatment of TMJ isorders: A systematic review of the literature. Cranio. 2010;**28**:166-176

[59] Iturriaga V, Bornhardt T, Manterola C, Brebi P. Effect of hyaluronic acid on the regulation of inflammatory

mediators in osteoarthritis of the temporomandibular joint: A systematic review. International Journal of Oral and Maxillofacial Surgery. 2017;**46**:590-595

[60] Andia I, Abate M. Platelet-rich plasma: Underlying biology and clinical correlates. Regenerative Medicine. 2013;**8**:645-658

[61] Andia I, Abate M. Platelet-rich plasma: Combinational treatment modalities for musculoskeletal conditions. Frontiers of Medicine. 2018;**12**:139-152

[62] Andia I, Maffulli N. A contemporary view of platelet-rich plasma therapies: Moving toward refined clinical protocols and precise indications. Regenerative Medicine. 2018;**13**:717-728

[63] Andia I, Maffulli N. Muscle and tendon injuries: The role of biological interventions to promote and assist healing and recovery. Arthroscopy. 2015;**31**:999-1015

[64] Tohidnezhad M, Bayer A, Rasuo B, Hock JVP, Kweider N, Fragoulis A, et al. Platelet-released growth factors modulate the secretion of cytokines in synoviocytes under inflammatory joint disease. Mediators of Inflammation. 2017;**2017**:1046438

[65] Chang KV, Hung CY, Aliwarga F, Wang TG, Han DS, Chen WS. Comparative effectiveness of platelet-rich plasma injections for treating knee joint cartilage degenerative pathology: A systematic review and meta-analysis. Archives of Physical Medicine and Rehabilitation. 2014;**95**:562-575

[66] Zubair U, Salam O, Zubair Z. Role of intra-articular platelet rich plasma in the management of osteoarthritis: A review. Cureus. 2018;**10**(9):e3359

[67] Kütük N, Baş B, Soylu E, Gönen ZB, Yilmaz C, Balcioğlu E, et al. Effect of platelet-rich plasma on fibrocartilage, cartilage, and bone repair in temporomandibular joint. Journal of Oral and Maxillofacial Surgery. 2014;**72**:277-284

[68] Hegab AF, Ali HE, Elmasry M, Khallaf MG. Platelet-rich plasma injection as an effective treatment for temporomandibular joint osteoarthritis. Journal of Oral and Maxillofacial Surgery. 2015;**73**:1706-1713

[69] Bousnaki M, Bakopoulou A, Koidis P. Platelet-rich plasma for the therapeutic management of temporomandibular joint disorders: A systematic review. International Journal of Oral and Maxillofacial Surgery. 2018;**47**:188-198

[70] Chung PY, Lin MT, Chang HP. Effectiveness of platelet-rich plasma injection in patients with temporomandibular joint osteoarthritis: A systematic review and meta-analysis of randomized controlled trials. Oral Surgery, Oral Medicine, Oral Pathology, Oral Radiology. 2018;**pii**:S2212-4403(18)31187-8

[71] Fernández Sanromán J, Fernández Ferro M, Costas López A, Arenaz Bua J, López A. Does injection of plasma rich in growth factors after temporomandibular joint arthroscopy improve outcomes in patients with Wilkes stage IV internal derangement? A randomized prospective clinical study. International Journal of Oral and Maxillofacial Surgery. 2016;**45**:828-835

[72] Fernández-Ferro M, Fernández-Sanromán J, Blanco-Carrión A, Costas-López A, López-Betancourt A, Arenaz-Bua J, et al. Comparison of intra-articular injection of plasma rich in growth factors versus hyaluronic acid following arthroscopy in the treatment of temporomandibular dysfunction: A randomised prospective study. Journal of Cranio-Maxillo-Facial Surgery. 2017;**45**:449-454

[73] Cömert Kiliç S, Güngörmüş M. Is arthrocentesis plus platelet-rich plasma superior to arthrocentesis plus hyaluronic acid for the treatment of temporomandibular joint osteoarthritis: A randomized clinical trial. International Journal of Oral and Maxillofacial Surgery. 2016;**45**:1538-1544

[74] Hancı M, Karamese M, Tosun Z, Aktan TM, Duman S, Savaci N. Intra-articular platelet-rich plasma injection for the treatment of temporomandibular disorders and a comparison with arthrocentesis. Journal of Cranio-Maxillo-Facial Surgery. 2015;**43**:162-166

[75] Almarza AJ, Brown BN, Arzi B, Ângelo DF, Chung W, Badylak SF, et al. Preclinical animal models for temporomandibular joint tissue engineering. Tissue Engineering. Part B, Reviews. 2018;**24**:171-178

[76] Shu W, Liu L, Bao G, Kang H. Tissue engineering of the temporomandibular joint disc: Current status and future trends. The International Journal of Artificial Organs. 2015;**38**:55-68

[77] Aryaei A, Vapniarsky N, JC1 H, Athanasiou KA. Recent tissue engineering advances for the treatment of temporomandibular joint disorders. Current Osteoporosis Reports. 2016;**14**:269-279

[78] Salash JR, Hossameldin RH, Almarza AJ. Potential indications for tissue engineering in temporomandibular joint surgery. Journal of Oral and Maxillofacial Surgery. 2016;(4):705-711

[79] Cui D, Li H, Xu X, Ye L, Zhou X, Zheng L, et al. Mesenchymal stem cells for cartilage regeneration of TMJ osteoarthritis. Stem Cells International. 2017;**2017**:5979741

[80] Acri TM, Shin K, Seol D, Laird NZ, Song I, Geary SM, et al. Tissue engineering for the temporomandibular joint. Advanced Healthcare Materials. 2018;**17**:e1801236

[81] Van Bellinghen X, Idoux-Gillet Y, Pugliano M, Strub M, Bornert F, Clauss F, et al. Temporomandibular joint regenerative medicine. International Journal of Molecular Sciences. 2018;**19**(2):pii: E446

[82] Kim IL, Mauck RL, Burdick JA. Hydrogel design for cartilage tissue engineering: A case study with hyaluronic acid. Biomaterials. 2011;**32**:8771-8782

[83] Willerth SM, Sakiyama-Elbert SE. Combining stem cells and biomaterial scaffolds for constructing tissues and cell delivery. In: StemBook, editor. The Stem Cell Research Community. Cambridge, MA, USA: Harvard Stem Cell Institute; 2008

[84] Bodugoz-Senturk H, Macias CE, Kung JH, Muratoglu OK. Poly(vinyl alcohol)-acrylamide hydrogels as load-bearing cartilage substitute. Biomaterials. 2009;**30**:589-596

[85] Uematsu K, Hattori K, Ishimoto Y, Yamauchi J, Habata T, Takakura Y, et al. Cartilage regeneration using mesenchymal stem cells and a three-dimensional poly-lactic-glycolic acid (PLGA) scaffold. Biomaterials. 2005;**26**:4273-4279

[86] Wang L, Lazebnik M, Detamore MS. Hyaline cartilage cells outperform mandibular condylar cartilage cells in a TMJ fibrocartilage tissue engineering application. Osteoarthritis and Cartilage. 2009;**17**:346-353

[87] Anderson DE, Athanasiou KA. A comparison of primary and passaged chondrocytes for use in engineering the temporomandibular joint. Archives of Oral Biology. 2009;**54**:138-145

[88] Wu Y, Gong Z, Li J, Meng Q, Fang W, Long X. The pilot study of fibrin

with temporomandibular joint derived synovial stem cells in repairing TMJ disc perforation. BioMed Research International. 2014;**2014**:454021

[89] Koyama N, Okubo Y, Nakao K, Osawa K, Fujimura K, Bessho K. Pluripotency of mesenchymal cells derived from synovial fluid in patients with temporomandibular joint disorder. Life Sciences. 2011;**89**:741-747

[90] Sunil P, Manikandhan R, Muthu M, Abraham S. Stem cell therapy in oral and maxillofacial region: An overview. Journal of Oral and Maxillofacial Pathology. 2012;**16**:58-63

[91] Mäenpää K, Ellä V, Mauno J, Kellomäki M, Suuronen R, Ylikomi T, et al. Use of adipose stem cells and polylactide discs for tissue engineering of the temporomandibular joint disc. Journal of the Royal Society Interface. 2010;**7**:177-188

[92] Guo L, Li J, Qiao X, Yu M, Tang W, Wang H, et al. Comparison of odontogenic differentiation of human dental follicle cells and human dental papilla cells. PLoS One. 2013;**8**:e62332

[93] Hagandora CK, Gao J, Wang Y, Almarza AJ. Poly(glycerol sebacate): A novel scaffold material for temporomandibular joint disc engineering. Tissue Engineering Part A. 2013;**19**:729-737

[94] Almarza AJ, Athanasiou KA. Seeding techniques and scaffolding choice for tissue engineering of the temporomandibular joint disk. Tissue Engineering. 2004;**10**:1787-1795

[95] Almarza AJ, Athanasiou KA. Effects of initial cell seeding density for the tissue engineering of the temporomandibular joint disc. Annals of Biomedical Engineering. 2005;**33**:943-950

[96] Allen KD, Athanasiou KA. Scaffold and growth factor selection in temporomandibular joint disc engineering. Journal of Dental Research. 2008;**87**:180-185

[97] Legemate K, Tarafder S, Jun Y, Lee CH. Engineering human TMJ discs with protein-releasing 3D-printed scaffolds. Journal of Dental Research. 2016;**95**:800-807

[98] Springer IN, Fleiner B, Jepsen S, Açil Y. Culture of cells gained from temporomandibular joint cartilage on non-absorbable scaffolds. Biomaterials. 2001;**22**:2569-2577

[99] Suh JK, Matthew HW. Application of chitosan-based polysaccharide biomaterials in cartilage tissue engineering: A review. Biomaterials. 2000;**21**:2589-2598

[100] Brown BN, Chung WL, Almarza AJ, Pavlick MD, Reppas SN, Ochs MW, et al. Inductive, scaffold-based, regenerative medicine approach to reconstruction of the temporomandibular joint disk. Journal of Oral and Maxillofacial Surgery. 2012;**70**:2656-2668

[101] Kinoshita Y, Maeda H. Recent developments of functional scaffolds for craniomaxillofacial bone tissue engineering applications. Scientific World Journal. 2013;**2013**:863157

[102] Zhang S, Yap AU, Toh WS. Stem cells for temporomandibular joint repair and regeneration. Stem Cell Reviews. 2015;**11**:728-742

[103] Sun YP, Zheng YH, Liu WJ, Zheng YL, Zhang ZG. Synovium fragment-derived cells exhibit characteristics similar to those of dissociated multipotent cells in synovial fluid of the temporomandibular joint. PLoS One. 2014;**9**:e101896

[104] Meirelles S, Fontes AM, Covas DT, Caplan AI. Mechanisms involved in the therapeutic properties of mesenchymal stem cells. Cytokine & Growth Factor Reviews. 2009;**20**:419-427

[105] Almarza AJ, Athanasiou KA. Evaluation of three growth factors in combinations of two for temporomandibular joint disc tissue engineering. Archives of Oral Biology. 2006;**51**:215-221

[106] Kalpakci KN, Kim EJ, Athanasiou KA. Assessment of growth factor treatment on fibrochondrocyte and chondrocyte co-cultures for TMJ fibrocartilage engineering. Acta Biomaterialia. 2011;**7**:1710-1718

[107] Kang H, Bi YD, Li ZQ, Qi MY, Peng EM. Effect of transforming growth factor β(1) and insulin-like growth factor-I on extracelluar matrix synthesis of self-assembled constructs of goat temporomandibular joint disc. Zhonghua Kou Qiang Yi Xue Za Zhi. 2011;**46**:541-546

[108] Jiao Y, Wang D, Han WL. Effects of various growth factors on human mandibular condylar cartilage cell proliferation. Zhonghua Kou Qiang Yi Xue Za Zhi. 2000;**35**:346-349

[109] Hanaoka K, Tanaka E, Takata T, Miyauchi M, Aoyama J, Kawai N, et al. Platelet-derived growth factor enhances proliferation and matrix synthesis of temporomandibular joint disc-derived cells. The Angle Orthodontist. 2006;**76**:486-492

[110] Ankrum J, Karp JM. Mesenchymal stem cell therapy: Two steps forward, one step back. Trends in Molecular Medicine. 2010;**16**:203-209

[111] Zhang J, Guo F, Mi J, Zhang Z. Periodontal ligament mesenchymal stromal cells increase proliferation and glycosaminoglycans formation of temporomandibular joint derived fibrochondrocytes. BioMed Research International. 2014;**2014**:410167

[112] Lai RC, Tan SS, Teh BJ, Sze SK, Arslan F, de Kleijn DP, et al. Proteolytic potential of the MSC exosome proteome: Implications for an exosome-mediated delivery of therapeutic proteasome. International Journal of Proteomics. 2012;**2012**:971907

Chapter 4

The Potential Effect of Medicinal Plants for Cartilage Regeneration

Franca Nneka Alaribe, Mapula Razwinani, Makwese Maepa and Keolebogile Shirley Caroline Motaung

Abstract

Any trauma to a joint such as sports injury can lead to osteoarthritis especially injuries that include torn cartilage, dislocated joints and ligaments. In sports injury specifically, most of the ointments in the market are only applied after physical activity. Repair of the bone and cartilage continues to be a challenge. Autologous and allografts are the gold standard for the treatment of the bone and cartilage. They have an invasive, open surgical procedure that requires the tissue to be harvested from an alternative site within the patient. South Africa is rich in native flora that is currently tapped as medicine by traditional healers. However, little is known about the natural products of our native flora and their potential to serve as a remedy for sports injuries, fracture healing and osteoarthritis. The grand purpose of the project is to explore medicinal plants of South Africa as a potential source for bone and tissue engineering of articular cartilage.

Keywords: tissue engineering, anti-inflammatory, osteoarthritis, sports injury, medicinal plants

1. Introduction

Generally, with the extensive screening of plants used in traditional medicine, evidence of their rational use in treating infections, diseases, inflammation and other disorders has been provided [1–3]. Herbal extracts have extensively health benefits, and indigenous medicinal plants have been used traditionally as a major source of drugs for the treatment of various illnesses, including osteoarthritis (OA), asthma, cancer, heart disease, tuberculosis, swollen ankles and hypertension [4–6]. Extracted compounds of medicinal plants are usually used as inputs in toxicology, phytochemicals, pharmaceuticals and other chemical industries [3–5, 7–9]. Stem cell therapies involving cartilage regeneration and several current 3D bioprinting processes involve the use of synthetic and natural biological molecules such as growth factors to improve their proliferation and differentiation [9–11]. There is an ongoing search in the science community for alternatives of these growth factors and the existing synthetic materials, due to reports on their numerous negative effects and complete failure in cartilage regeneration [3, 12–14]. Several medicinal plant extracts have been suggested to stimulate adult stem cell proliferation and thus regeneration of damaged or diseased tissues. Many Chinese herbs have been found to exert adipogenic, osteogenic and chondrogenic effects on human mesenchymal stem cells (hMSCs). Dried root of *Drynaria fortunei* contains flavonoid and triterpenoid found to promote increased bone cell viability, intracellular total protein as well as alkaline and acid phosphates. Naringin, the major component of *Rhizoma drynariae* extract, enhanced the proliferation of BM-derived hMSCs by regulating β-catenin and AMP-activated protein kinase (AMPK) [15–17]. *Foeniculum vulgare* is traditionally used in the estrogenic activity to enhance milk secretion, in birth facilitation and for the alleviation of dysmenorrhea. *Foeniculum vulgare* extract has been found to promote the proliferation and differentiation of BM-derived hMSC into osteoblasts. Additionally, an ethanol extract of *Ferula gummosa* (an Iranian traditional medicine) was observed to enhance proliferation and differentiation of BM-derived hMSCs into osteocytes. [18]. Studies in this section elaborate on the possible mechanisms and beneficial effects of herbal remedies in the engineering of articular cartilage and regenerative medicine.

1.1 Role of medicinal plants in chondrocytes

In South Africa, numerous plants used traditionally have been employed in tissue engineering of articular cartilage. Studies have observed medicinal plants such as *Pleurostylia capensis*, *Pterocarpus angolensis* and *Eucomis autumnalis*, having resveratrol playing proliferation and differentiation roles in tissue engineering of articular cartilage. High regulation of collagen type II has been observed chondrocytes treated with resveratrol [19]. This makes resveratrol potentially enhancing chondrocyte viability which can be applied in 3D bioprinting of cartilage constructs [20]. Recent publications show that bark and root water extracts of *Pterocarpus angolensis* plants in the stifle joints from the 3-month-old pig affect the accumulation of collagen type II in porcine articular cartilage in the middle zone. Cell culture experiments were designed to investigate the role of the bark and root water extracts of *P. angolensis* to induce the expression of collagen type II protein in porcine articular chondrocytes. Monolayer cells were treated with 15, 30 and 50 μg/ml of *P. angolensis* extract and hydrogen peroxide (2 μg/ml) for 4 days, and the untreated chondrocytes were used as controls. The results showed no significant difference in the cell index between the controls and chondrocytes that had been treated with the plant extracts at 15 and 30 μg/ml. A significant increase in the expression of collagen type II protein by the chondrocytes was observed and found to be optimal at a concentration of 30 μg/ml. There was an increase in the production of proteoglycans. However, the plant extracts at a 50 μg/ml induced apoptosis in the middle zone chondrocytes. In conclusion the findings of this study are of great importance in understanding the mechanisms through which *P. angolensis* enables the healing of breached tissue [21]. In our laboratory, an (unpublished) in vitro study has observed the enhancement of proliferation and osteogenic differentiation (by increasing alkaline phosphate activity) of C2C12 myoblast cells treated with *Pleurostylia capensis* crude extract. Furthermore, proliferation and lineage differentiation of *P. angolensis* and *E. autumnalis* in porcine adipose-derived mesenchymal stem cells (pADMSCs) have also been recorded in our work (Figure not shown). However, the potential use of medicinal plants with tissue engineering methods to treat the cartilage and bone is exciting, yet not fully realized, and is likely to be a future treatment strategy.

1.2 Medicinal plant extract in scaffolds

Signals, morphogens responding stem cells and scaffolds that are biomimetic of the extracellular matrix are the three paramount requirements in regenerative medicine [22, 23]. Currently, empirical formulations, medicinal plants and their bioactive compounds are being merged with polymers that can be used in tissue regeneration.

Many studies have tried to incorporate medicinal plants in the fabrication of different scaffolds for wound healing, bone fracture and cartilage regeneration.

Herbal plants have the potential in tissue engineering and regenerative medicine due to their minimal host inflammatory response, high level of tenability and the ability to progressively degenerate into non-cytotoxic components, which are either reabsorbed or removed from the biological system [24]. Recently, studies have shown that scaffolds treated with *Cissus quadrangularis* extract (known as Asthisandhani in Indian traditional medicine) exhibited significant differences with regard to hMSC proliferation, attachment and enhanced osteoblast differentiation properties compared with scaffolds not treated with the extract [25]. Young et al. [26] also incorporated *Terminalia bellirica* extract in a hydrogel composition for use in stem cell therapy. This extract was found to result in significantly higher rates of hMSC proliferation and cell attachment.

Similarly, we have evaluated natural polymer (chitosan and alginate) scaffolds incorporated with *E autumnalis* and *P. angolensis* extracts as done in [25] on pADMSCs for lineage differentiation. The attachment capacity was evaluated by incubating pADMSCs with herbal and non-herbal scaffold at different concentrations of 1, 3 and 5 mg/ml. The samples were further stained with 4′,6-diamidino-2-phenylindole and calcein green after 72 h according to the manufacturer's instructions. The pADMSCs incubated with herbal scaffolds showed significant differences with regard to proliferation and cell attachment compared to pADMSCs incubated with non-herbal scaffolds (**Figure 1**). A higher number of cells were obviously present and attached to the herbal scaffolds in DAPI staining (**Figure 1d** and **e**) than in non-herbal scaffold (**Figure 1f**). A similar condition was also observed in the calcein staining with herbal scaffold enhancing cell proliferation and attachment (**Figure 1g** and **h**) compared to in the non-herbal scaffold (**Figure 1i**).

The chondrogenic differentiation capacity of the herbal scaffolds was also evaluated using toluidine blue staining after 21 days in culture (**Figure 2**). Herbal scaffolds were found to enhance formation of chondrocytes (**Figure 2a** and **c**) compared to non-herbal scaffolds (**Figure 2e**). Herbal scaffolds also showed significant chondrogenic enhancement compared to the controls (**Figure 2b**, **d** and **f**).

Additionally, our anti-inflammatory assay for days 7, 14 and 21 using an interleukin 6 (IL-6) Elisa kit according to the manufacturer's instructions confirmed the anti-inflammatory nature of *E. autumnalis* and *P. angolensis*. Inflammation was significantly higher ($P < 0.01$) in cells cultured with interleukin 6 and non-herbal scaffolds than in herbal scaffolds. The herbal scaffolds suppressed the expression of IL-6 in the cultured pADMSCs.

The degradation of mineral deposition during the in vitro regeneration process in tissue engineering is very important. Hence, we tried to assess the in vitro degradation capacity of our scaffolds using scanning electron microscope (SEM). It was observed that our herbal scaffolds showed significantly higher and gradual releasing of materials into the culture environment than our non-herbal scaffolds. The in vitro mineral deposition was confirmed using Fourier transform infrared spectrometer (FT-IR) spectrum (**Figure 3a** and **b**) on day 14 of incubation with pADMSCs in culture. The FT-IR data for the herbal scaffolds (**Figure 3a**) has an open-chain bond —C=N— at peaks 1600.8 and 1416.4 which were reduced. The 1072.4 and 1029.8 peaks were longer and more pronounced. The peak bands after 824.15 that are assumed to be vibrations of P—O—H from $Ca_3(PO_4)_2$ seems to be extended to peak 450. In the case of the non-herbal scaffolds (**Figure 3b**), peak bands at 1600.8 and 1416.4 were longer and seen at 1593.7 and 1420, respectively. At 1072.4 it is almost absent and the peak band at 1015.7 is reduced.

The FT-IR analysis showed certain peaks which are in the same functional groups as alkyl carbonate, organic sulphate and phosphate ions [27, 28].

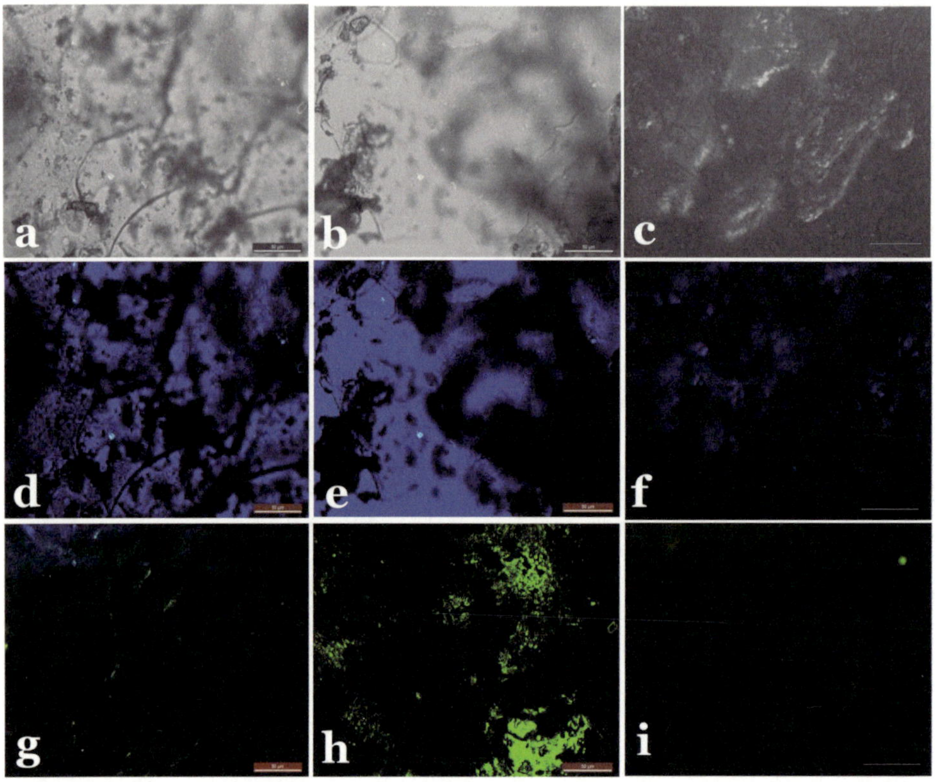

Figure 1.
Immunofluorescence staining of scaffolds (3 mg/ml) cultured for 72 h in adipose-derived porcine mesenchymal stem cells. (a–c) Images recorded under white field, (d–f) DAPI stain, (g–i) calcein stain, (a, d, g) Eucomis autumnalis scaffold, (b, e, h) Pterocarpus angolensis scaffold, and (c, f, i) cells cultured in non-herbal scaffold. Scale bar, 50 μm; magnification, 10×.

Furthermore, the presence of calcium, phosphate and carbonate compounds highlights the important relationship between intracellular calcium phosphate in osteoblasts and their role in mineralizing the extracellular matrix [29]. The long sharp peak at 1017.2 cm^{-1} also corresponds to silicate (Si) ions. Silicate and Cu ions are usually encountered in the presence of a hydrated surface layer of both bone crystal and synthetic apatite crystals, which contain varying concentrations of a wide variety of mineral ions that play important roles during bone and cartilage regeneration [30].

1.3 Medicinal plant extracts in wound healing

The skin is susceptible to injury and is the body tissue most exposed to damage. Wound healing is a normal biological process involving proliferation and redifferentiation of fibroblasts and keratinocytes [31, 32]. Significant advances have been made in the past years in wound healing so as to bring solutions for the treatment of chronic wounds and speeding up of acute healing. Several recent studies have found plants to be significant in controlling wound healing [33, 34].

Scrophularia striata, a well-known plant in Iranian traditional medicine, has anti-oxidative and anti-inflammatory properties. It is traditionally employed in wound healing due to these mentioned properties. Ghashghaii et al. [35] evaluated the wound healing potential of *S. striata* on cutaneous wounds in rats. Data from the study showed that rats treated with *S. striata* showed a significant decrease in the wound area, with a decrease in the number of lymphocytes,

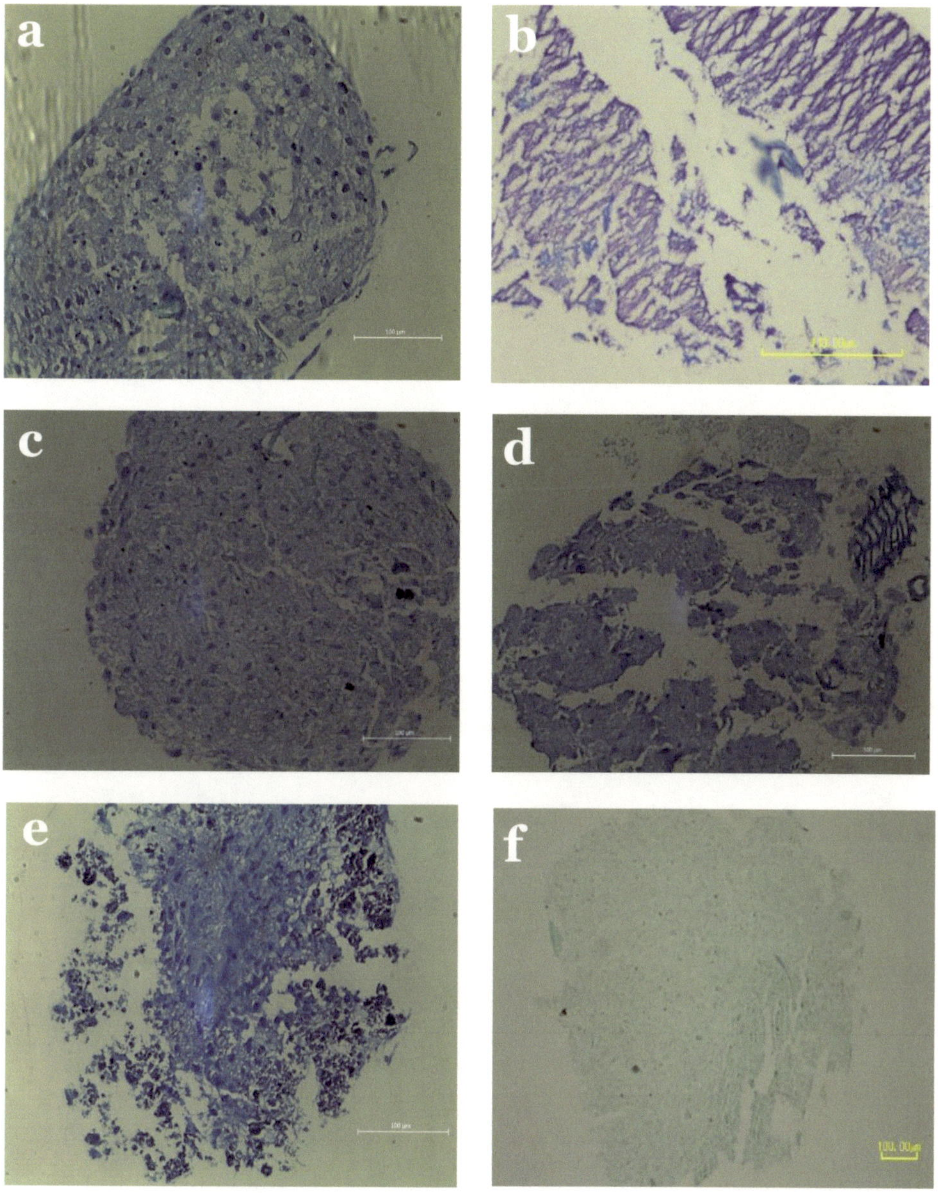

Figure 2.
Chondrogenic evaluation of the three experimental groups of the scaffold (3 mg/ml) with pADMSCs micro-mass pellet stained with toluidine blue at day 21 of treatment. (a) E. autumnalis herbal scaffold, (b) transforming growth factor (TGF)-beta 1 10 ng/ml positive control, (c) P. angolensis scaffold, (d) bone morphogenetic protein-2 (BMP-2, 10 ng/ml), (e) non-herbal scaffold and (f) negative control, pADMSCs without treatment. Scale bar, 100 μm; magnification, 10×.

enhanced number of fibroblasts and epithelial formation that resulted to early maturity of the collagen fibres compared to other groups. The study generally showed that application of *S. striata* on wounds resulted in substantial contraction and faster wound healing, which makes *S. striata* a potential subject for the treatment of wounds in animals and human beings.

Additionally, *Anogeissus leiocarpus*, a Ghanaian traditional plant, has been evaluated for wound healing activities in albino Wistar rats. A study of the wound healing effect of *A. leiocarpus* extract gave an interesting result. The plant formulation showed a progressive decrease in wound area with time [36, 37]. At day 15, the

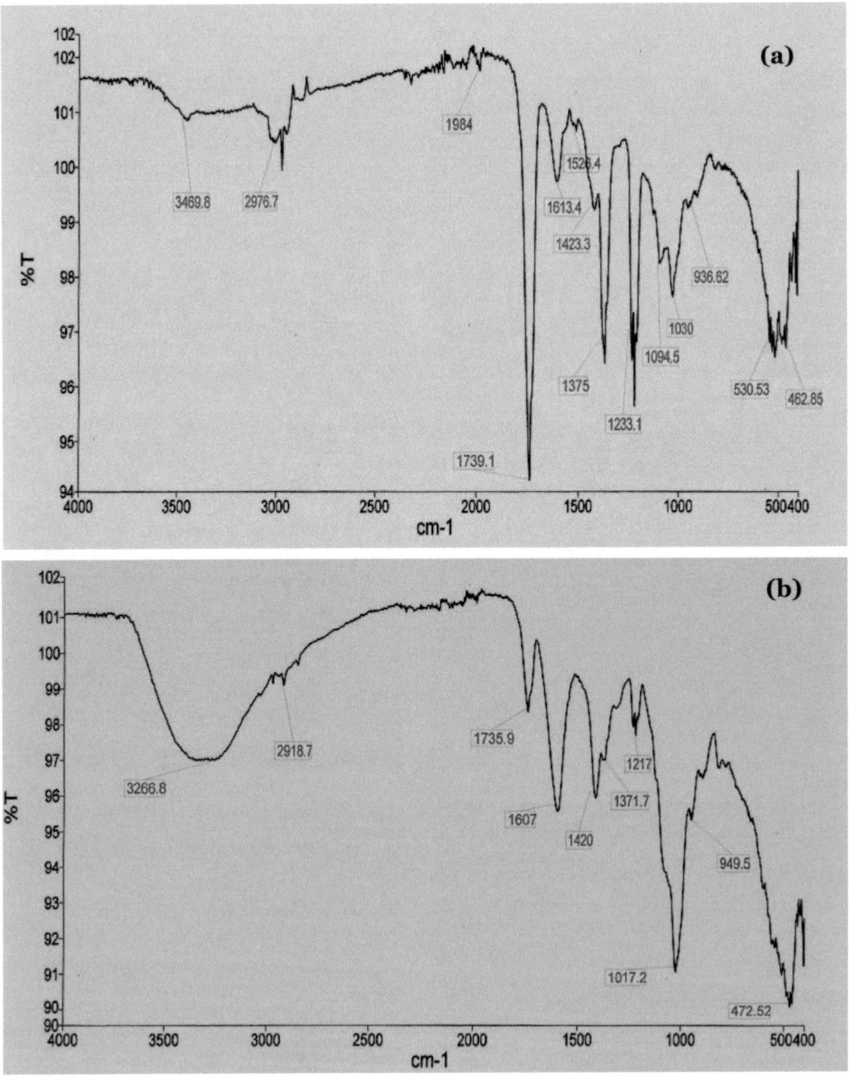

Figure 3.
The FT-IR spectrum of the scaffolds in culture with pADMSC cells at day 14 to confirm biomineralization. The analysis was done using the KBr method in the range of 400–4000 cm^{-1}*. (a) Herbal scaffold and (b) non-herbal scaffold.*

mixture containing 100 mg/ml aqueous extract and 10% w/w powdered ointment of *A. leiocarpus* showed 100% healing similar to the standard antibiotic (2% w/w penicillin).

Furthermore, a study has used *Moringa* extract incorporated with nanofibrous polyacrylonitrile for wound healing. Data from the study showed that *Moringa* influenced the healing properties of the material. At days 1, 4 and 7 of the wound dressing experiment, the percentage wound closure of the rat was the highest for the nanofiber containing 0.5 g of *Moringa* leaf extract (35, 87 and 95%, respectively) compared to the positive control medical gauze (29, 75 and 93%, respectively) [38].

Similarly, our study also evaluated the wound healing capacity of *E. autumnalis* and *P. angolensis* using the subcutaneous porcine adipose-derived stem cells up to 72 h, as done in [39, 40] with a slight modification. Percentage wound healing closure was calculated using the equation: initial area of wound—nth day area of wound/

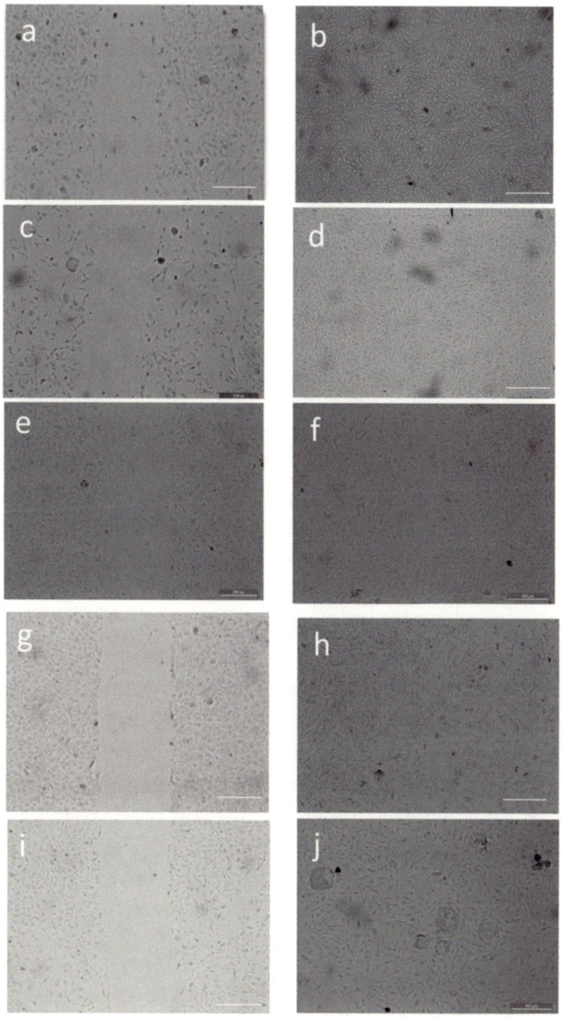

A

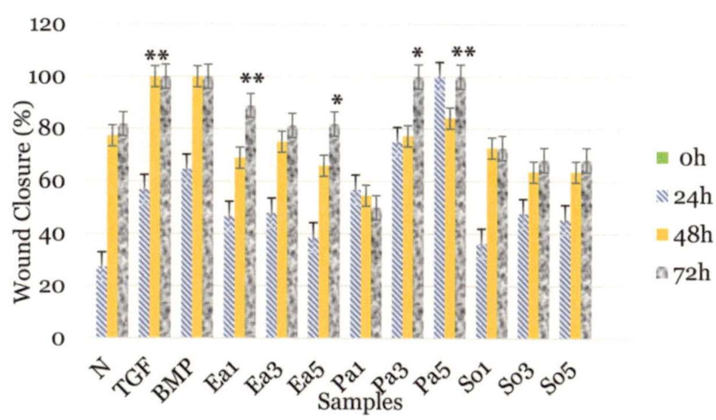

B

Figure 4.
*In vitro wound closure appearance of the adipose subcutaneous cells after treatment with herbal and non-herbal scaffold media at 5 mg/ml. (A) 0 h (a, c, e, g and i) and 72 h (b, d, f, h and j). (a–b) Scaffold with E. autumnalis extract; (c–d) scaffold with P. angolensis extract; (e–f) negative control, scaffold without extract; (g–h) positive control, TGF 10 ng/ml; and (i–j) positive control, BMP-2 10 ng/ml. Scale bar, 100 μm; magnification, 10×. (B) Wound healing percentage (%) at 0, 24, 48 and 72 h of treatment with 1, 3 and 5 mg/ml of E. autumnalis and P. angolensis extract scaffold media. The data are expressed as mean ± standard deviation from six independent experiments, ** ($p < 0.01$) and * ($p < 0.05$).*

initial area of wound × 100. Data from our in vitro study (**Figure 4**) showed that the herbal extracts influenced the in vitro healing capacity of the cellulose/alginate polymer scaffolds. The healing capacity was found to be significantly higher ($P < 0.01$) in *P. angolensis* (**Figure 4A (a–b)**, **B** Pa3 and Pa5) at 24 and 72 h, respectively, compared with the non-herbal scaffold (**Figure 4A (e–f)**, **B** So3 and S05). The *E. autumnalis* extract performed well and was statistically significant (**Figure 4A (c–d)**, **B** Ea1 ($P < 0.01$) and Ea5 ($P < 0.05$)) at 72 h, respectively, but was low compared to the positive controls (**Figure 4A (g–j)**, **B** TGF ($P < 0.01$) and BMP-2, respectively). Our data so far depicted that herbal extracts improved the wound healing capacity with the incorporated natural biopolymers.

2. Conclusion

Numerous polymeric constructs have been used in combination with growth factors for engineering and regeneration of tissues. This combination of polymer and growth factors for tissue repair depends largely on using biodegradable materials that can stimulate specific cellular responses at a molecular level which should be suitable, simple and cost-effective. Our data in this section offers pharmacological evidence on the potential use of the mentioned plant extracts in bone fracture, cartilage regeneration and wound treatment. In fact, medicinal plants found to have anti-inflammatory properties may partake in host modulatory therapy for various inflammatory diseases as proposed in [3, 39].

We would like to state that the herbs and all the substances in this study are for cartilage defects of grades 1, 2 and 3 according to outerbridge scale. Therefore, if congenital or after trauma large cartilage case is presented, then operative treatment is advised.

Acknowledgements

This research study was supported by the South African National Research Foundation. We thank the Tshwane University of Technology for their institutional support and Vleis 1 abattoir for providing pig knees.

Author details

Franca Nneka Alaribe, Mapula Razwinani, Makwese Maepa and Keolebogile Shirley Caroline Motaung*
Department of Biomedical Sciences, Faculty of Science, Tshwane University of Technology, Pretoria, South Africa

*Address all correspondence to: smotaung@gmail.com

References

[1] Palhares RM, Gonçalves DM, Dos Santos AFB, Pereira CG, das Graças LBM, Oliveira G. Medicinal plants recommended by the World Health Organization: DNA barcode identification associated with chemical analyses guarantees their quality. PLoS One. 2015;**10**(5):e0127866

[2] Dar RA, Shahnawaz M, Qazi PH. General overview of medicinal plants: A review. The Journal of Phytopharmacology. 2017;**6**(6):349-351

[3] Alaribe FN, Maepa MJ, Mkhumbeni N, Motaung SCKM. Possible roles of *Eucomis autumnalis* in bone and cartilage regeneration: A review. Tropical Journal of Pharmaceutical Research. 2018;**17**(4):741-749

[4] Bisi-Johnson MA, Obi CL, Hattori T, Oshima Y, Li S, Kambizi L, et al. Evaluation of the antibacterial and anticancer activities of some South African medicinal plants. BMC Complementary and Alternative Medicine. 2011;**11**:14

[5] Koorbanally C, Crouch NR, Mulholland DA. The phytochemistry and ethnobotany of the Southern African genus *Eucomis* (Hyacinthaceae: Hyacinthoideae). Phytochemistry: Advances in Research. 2006:69-85

[6] Jalil AM, Shuid AN, Muhammad N. Role of medicinal plants and natural products on osteoporotic fracture healing. Evidence-Based Complementary and Alternative Medicine. 2012:714512

[7] O'Donnell S, Lagacé C, McRae L, Bancej C. Life with arthritis in Canada: A personal and public health challenge. Chronic Diseases and Injuries in Canada. 2011;**31**(3):135-136

[8] Abere TA, Okoto PE, Agoreyo FO. Antidiarrhoea and toxicological evaluation of the leaf extract of dissotis rotundifolia triana (melastomataceae). BMC Complementary and Alternative Medicine. 2010;**10**:71

[9] Dzobo K, Thormford NE, Senthebane DA, et al. Advances in regenerative medicine and tissue engineering: Innovation and transformation of medicine. Stem Cells International. 2018:101155-2495848

[10] Nowicki M, Wierzbowska A, Malachowski R, et al. VEGF, ANGPT1, ANGPT2, and MMP-9 expression in the autologous hematopoietic stem cell transplantation and its impact on the time to engraftment. Annals of Hematology. 2017;**96**(12):2103-2112

[11] Eichler F, Duncan C, Musolino PL, et al. Hematopoietic stem-cell gene therapy for cerebral adrenoleukodystrophy. The New England Journal of Medicine. 2017;**377**(17):1630-1638

[12] Sarvazyan A et al. Muscle as a molecular machine for protecting joints and bones by absorbing mechanical impacts. Medical Hypotheses. 2014;**83**(1):6-10

[13] Cai S, Han L, Ao Q, Chan YS, Shum DKY. Human induced pluripotent cell-derived sensory neurons for fate commitment of bone marrow-derived Schwann cells: Implications for remyelination therapy. Stem Cells Translational Medicine. 2017;**6**(2):369-381

[14] Huizinga TWJ, Gröndal G. Drivers of costly treatment strategies in rheumatoid arthritis. The Lancet. 2016;**388**(10041):213-214

[15] Zhang P, Dai KR, Yan SG, Yan WQ, Zhang C, Chen DQ, et al. Effects of naringin on the proliferation and osteogenic differentiation of

human bone mesenchymal stem cell. European Journal of Pharmacology. 2009;**607**(1-3):1-5

[16] Wang D, Ma W, Wang F, Dong J, Wang D, Sun B, et al. Stimulation of Wnt/β-catenin signaling to improve bone development by naringin via interacting with AMPK and Akt. Cellular Physiology and Biochemistry. 2015;**36**(4):1563-1576

[17] Udalamaththa DL, Jayasinghe CD, Udagama PV. Potential role of herbal remedies in stem cell therapy: Proliferation and differentiation of human mesenchymal stromal cells. Stem Cell Research & Therapy. 2016;**7**:110. DOI: 10.1186/s13287-016-0366-4

[18] Mahmoudi Z, Soleimani FM, Saidi A, Iranshahi M, Azizsoltanli A. Effect of Ferula gummosa ethanolic extract on osteogenesis in human mesenchymal stem cells. Journal of Medicinal Plants. 2013;**12**(46):50-59

[19] Mmadira MG, Motaung SCKM. The use of bone morphogenetic protein-7 and resveratrol in collagen type II of articular cartilage. Journal of Pharmacy and Pharmacology. 2016:199-211

[20] Maepa M, Razwinani M, Motaung S. Effects of resveratrol on collagen type ii protein in the superficial and middle zone chondrocytes of porcine articular cartilage. Journal of Ethnopharmacology. 2016;**178**:25-33

[21] Ssemakalu C, Razwinani M, Maepa MJ, Motaung KSCM. Pterocarpus angolensis crude extracts induce the expression of collagen type II in articular cartilage *in vitro*. African Journal of Traditional, Complementary, and Alternative Medicines. 2016;**13**(1):76-84

[22] Alaribe FN, Manoto SL, Motaung SCKM. Scaffolds from biomaterials: Advantages and limitations in bone and tissue engineering. Journal of Biologia. 2016;**71**(4)

[23] Motaung CKM, Cesare PE, Reddi H. Differential response of cartilage oligomeric matrix protein (COMP) to morphogens of bone morphogenetic protein/transforming growth factor-β family in the surface, middle and deep zones of articular cartilage. Journal of Tissue Engineering and Regenerative Medicine. 2011;**5**:87-96

[24] Martin JR, Gupta MK, Page JM, et al. A porous tissue engineering scaffold selectively degraded by cell-generated reactive oxygen species. Biomaterials. 2014;**35**(12):3766-3776

[25] Soumya S, Sajesh KM, Jayakumar R, Nair SV, Chennazhi KP. Development of a phytochemical scaffold for bone tissue engineering using Cissus quadrangularis extract. Carbohydrate Polymers. 2012;**87**(2):1787-1795

[26] Young YJ, Dolly S, Deepti S, Mi LE, Soonmo C, Soo HS, et al. *Terminalia bellirica* extracts loaded on stimuli responsive HEMA-DEA hydrogel for enhanced growth and proliferation of mesenchymal stem cells. Journal of Biomaterials and Tissue Engineering. 2014;**4**(1):37-45

[27] Coats J. Inteerpretation of infrared spectrum, a practical approach. In: Encyclopedia of Analytical Chemistry. Chichester: John Wiley & Sons Ltd.; 2000. pp. 10815-10837

[28] Jia J, Zhou H, We J, Jiang X, Hua H, Chen F, et al. Development of magnesium calcium phosphate biocement for bone regeneration. Journal of The Royal Society Interface. 2010;**7**:1171-1180

[29] Boonrungsiman S, Gentleman E, Carzaniga R, Evans ND, McComb DW, Porter AE, et al. The role of intracellular calcium phosphate in osteoblast-mediated bone apatite formation. PNAS. 2012;**109**(35):14170-14175

[30] Rey C, Combes C, Drouet C, Glimcher MJ, et al. Bone mineral: Update on chemical composition and structure. Osteoporosis International. 2009;**20**(6):1013-1021

[31] Boateng J, Catanzano O. Advanced therapeutic dressings for effective wound healing—A review. Journal of Pharmaceutical Sciences. 2015;**104**(11):3653-3680

[32] Gurtner GC, Werner S, Barrandon Y, Longaker MT. Wound repair and regeneration. Nature. 2008;**453**:314-321

[33] Rhoads DD, Cox SB, Rees EJ, Sun Y, Wolcott RD. Clinical identification of bacteria in human chronic wound infections: Culturing vs. 16S ribosomal DNA sequencing. BMC Infectious Diseases. 2012;**12**:321

[34] Jarić S, Mačukanović-Jocić M, Djurdjević L, Mitrović M, Kostić O, Karadžić B, et al. An ethnobotanical survey of traditionally used plants on Suva planina mountain (south-eastern Serbia). Journal of Ethnopharmacology. 2015;**175**:93-108

[35] Ghashghaii A, Hashemnia M, Nikousefat Z, Zangeneh MM, Zangeneh A. Wound healing potential of methanolic extract of scrophularia striata in rats. Pharmaceutical Sciences. 2017;**23**:256-263. DOI: 10.15171/PS.2017.38

[36] Barku VYA, Boye A, Ayaba S. Phytochemical screening and assessment of wound healing activity of the leaves of *Anogeissus leiocarpus*. European Journal of Experimental Biology; 2013;**3**(4):18-25

[37] Barros L, Ferreira MJ, Queirós B, Ferreira ICFR, Baptista P. Total phenols, ascorbic acid, β-carotene and lycopene in Portuguese wild edible mushrooms and their antioxidant activities. Food Chemistry. 2007, 2013;**103**:413-419

[38] Fayemi OE, Ekennia AE, Katata-Seru L, Ebokaiwe AP, et al. Antimicrobial and wound healing properties of polyacrylonitrile-moringa extract nanofibers. ACS Omega. 2018;**3**:4791-4797

[39] Singhal A, Gupta H, Bhati V. Wound healing activity of *Argyreia nervosa* leaves extract. International Journal of Applied & Basic Medical Research. 2011;**1**(1):36-39

[40] Nicolaus C, Junghanns S, Hartmann A, Murillo R, Ganzera M. *In vitro* studies to evaluate the wound healing properties of Calendula officinalis extracts. Journal of Ethnopharmacology. 2017;**196**:94-103

Chapter 5

Reconstruction with Joint Preservation

Lourenço Galizia Heitzmann

Abstract

The joint injury is a common disorder. Some techniques have been employed to repair the joint or regenerate the cartilage defects with different degrees of success. Four commonly performed techniques to preserve the joint included osteotomies, bone marrow stimulation, cartilage repair, and cartilage regeneration.

Keywords: cartilage, articular/injuries, cartilage, articular/surgery, chondrocytes/transplantation, periosteum/transplantation, treatment outcome

1. Introduction

Musculoskeletal injuries that disrupt the structure and function of diarthrodial joints can cause permanent biomechanical alterations and lead to a more severe, chronic condition. Despite advancements that have been made of restore tissue function and delay the need for joint replacement, there are currently no disease-modifying therapies for osteoarthritis (OA). To reduce the risk of OA, innovative preventive medicine approaches have been developed over the last decade to treat the underlying pathology.

The lesions of the articular cartilage are a common disorder that with the aging of the population its prevalence is increasing. More than 500,000 procedures are performed for the treatment of articular cartilage-related injuries, and many of these procedures are repeated in the same patients. This demonstrates the ineffectiveness of this isolated procedure [1].

Hunter [2] noted that the cartilage, "once destroyed, is not repaired." Fact that keeps current. Some studies have shown an incidence of cartilage lesions greater than 65% in routine arthroscopy [3–6]. No procedure nowadays is reliable for the regeneration of articular cartilage. This is due to the complexity of its structure and functional properties, such as minimizing friction and increasing the contact surface area to decrease wear under load.

Cartilage lesions (9 mm or greater) have been reported to be biomechanically unstable with a high propensity of progression to degenerative osteoarthritis [7, 8]. The main characteristics of the clinical presentation are pain, loss of movement, and alteration of function. Various surgical procedure options can be used for treatment; this will depend on the location, size, and stage [9, 10].

Articular cartilage is composed of chondrocytes (5–10%), water (65–80%), collagen, smaller glycoproteins such as fibronectin and oligomeric cartilage proteins, and large negatively charged hydrophilic proteoglycans (aggrecan, hyaluronan). Four distinct zones are described microscopically.

The superficial zone protects the deeper layers of shear stresses and composes approximately 10–20% of the thickness of the articular cartilage. The main collagen fibers found are type II and IX with a high number of flat chondrocytes. It is the layer that protects and maintains the integrity of the deeper layers, is in direct contact with the synovial fluid, and is responsible for most of the traction properties of the cartilage, which allows it to resist the pure, elastic, and compressive forces imposed by the joint.

The intermediate (transitional) zone provides an anatomic and functional bridge between the superficial and deep zones, accounts for 40–60% of the total cartilage volume, and contains thicker collagen proteoglycans and fibrils. Collagen is organized obliquely, and the chondrocytes are spherical and low density. The compressive forces mainly exhibit resistance.

The deep zone is responsible for providing greater resistance to the compressive forces, since the collagen fibrils are arranged perpendicular to the articular surface. The deep zone contains collagen fibrils of larger diameter in radial arrangement, higher content of proteoglycans, and lower concentration of water. Chondrocytes are typically arranged in columnar orientation, parallel to collagen fibers, and perpendicular to the joint line. It represents approximately 30% of the articular cartilage volume.

The calcified layer plays an integral role in the attachment of the cartilage to the bone, anchoring the collagen fibers from the deep zone to the subchondral bone. There are few cells and the chondrocytes are hypertrophic [9, 11–13].

Several factors are part of the etiology of the chondral or osteochondral lesion; among them are metabolic, such as obesity, alcohol abuse, and diabetes, as well as mechanical factors such as trauma, joint misalignment, and instability [12, 13].

Osteochondral lesions heal by formation of fibrocartilage secondary to the initial inflammatory response. Although mesenchymal cells produce collagen type I and II, the repair is mostly fibrocartilaginous in nature. The orderly structural organization of normal hyaline cartilage is lacking and results in early degradation and fragmentation. However, pure chondral lesions are painless and poorly repaired due to lack of vascularity [9, 14].

Surgical and nonoperative procedures are employed in the treatment of cartilage lesions. The main objective goal is to reduce pain and restore function. Nonsurgical treatments include physical therapy, activity modification, braces and orthoses, weight loss, steroid injections, chondroitin sulfate, and viscosupplementation with hyaluronic preparations [15–20]. The operative treatment aims to improve joint function and congruence as well as prevent osteoarthritic damage in intact areas of cartilage. It may be divided into three techniques commonly performed to preserve the joint including bone marrow stimulation, cartilage repair, and cartilage regeneration.

2. Bone marrow stimulation (BMS) techniques

2.1 Drilling/microfracture/abrasion techniques

Burmann in 1931, Haggart in 1940, and Magnuson in 1941 described joint debridement techniques for the treatment of osteoarthritis. Pridie in 1958 introduced the technique of perforation of the subchondral tissue exposing the vascularization of bone marrow, and later Ficat in 1979 described the spongialization, a resection of the entire subchondral bone plate chondromalacia patellae, with good to excellent results. Steadman suggested that specially designed awls are used to make multiple perforations or "microfractures," into the subchondral bone plate [21–30]. The perforations are made as close together as necessary, but not so close that one breaks into another. Consequently, the microfracture holes are approximately 3–4 mm apart (or three to four holes per square centimeter) [31, 32].

Chondroplasty by abrasion depends on the mechanical stimulation, like burrs, of the joint defect, without penetration of the subchondral bone. Exposure of small blood vessels generates formation in a clot attached to the surface. Fibrous tissue metaplasia occurs for fibrocartilage.

Multiple perforations have the benefit of causing less thermal damage than chondroplasty by abrasion and also leave the subchondral surface more rugged, allowing better adhesion of the blood clots. The penetration of the subchondral bone stimulates the local release of growth factors from the underlying bone. These factors attract and aid the differentiation of mesenchymal stem cells from the bone marrow in chondrocyte-like cells [33, 34].

Patients require a period of 6–8 weeks of non-weight-bearing to allow maturation of the fibrocartilage. Also, according to some authors, continuous passive motion for pain control and better function may be necessary [35].

For better results, some important factors include a body mass index below 30 kg/m^2, age under 40 years, defect less than 4 cm^2, volume of repaired cartilage (defective filling) greater than 66%, and symptoms less than 12 months [35].

The repair tissue may be able to fill the defect, but it lacks the normal histological or biomechanical properties of hyaline cartilage. Therefore, it has a stability inferior to the compressive and shear forces and tends to deteriorate with the time [35–39].

However, in their 2017 study, Frehner et al. concluded that treatment of osteochondral lesion by microfracture cannot be seen as an evidence-based procedure [39].

3. Cartilage replacement techniques

3.1 Chondrocyte autograft transfer and mosaicplasty

The description of the technique using osteochondral autografts for the treatment of joint defects was firstly studied by Pap and Krompecher [40]. Later, Wagner and Muller in Germany used the posterior part of the femoral condyle as an osteochondral autograft [41, 42]. Motions came in the 1990s by Matsusue in Japan and Hangody and colleagues in Hungary [43, 44].

The osteochondral plugs are harvested from non-weight-bearing areas and are transplanted into a small osteochondral defect. A larger lesion is filled in with multiple cylinders; it is also possible to transfer the posterior femoral condyle. Due to multiple cylinders, the gaps between the plugs produce an irregular articular surface.

The main indications for mosaicplasty include the chondral or focal osteochondral lesion in a stable knee, with lesions smaller than 22 mm in diameter and no more than 10 mm in depth.

The main benefits of this technique are that it is a single-stage procedure and there is rapid subchondral bone healing with restoration of native type II hyaline cartilage at the articular surface.

In a series by Hangody et al. with 57 patients and follow-up of more than 3 years, reported 91% good to excellent results with a mosaicplasty [45]. Gudas et al. in a prospective randomized study showed better clinical-functional and MRI results after 3 years for osteochondral transplants than for microfracture surgery [46].

Most of the studies showed good to excellent results in the short and long term, with a greater return to athletic activity when compared to microfracture [47–54].

Major complications of the osteochondral graft include donor site morbidity such as patellofemoral arthritis, fibrocartilage hypertrophy of the donor area, and unsatisfactory filling of the cartilage defect (especially with grafts > 8 mm in diameter) [49, 52, 54–56].

3.2 Osteoperiosteal graft

Another option we have for the treatment of osteochondral lesions is the mosaicplasty technique with bone-periosteum graft of the iliac crest. Its two advantages compared to the conventional technique include the absence of joint morbidity from the donor site defect [54] and that the periosteum (with its pluripotent stem cells) has the potential to differentiate into fibrocartilage [56–63].

3.2.1 Reconstruction with periosteal-cortical graft of the tibial lateral plateau: clinical outcome with an 18-month follow-up case

A 63-year-old woman presented with a bicondylar fracture of the right tibial plateau with extension to the diaphysis. She underwent surgical treatment 40 days after the fracture. It presented great destruction of the lateral articular surface, being reconstructed with the use of periosteal cortical graft of the external iliac board, suture of lateral meniscus, and reinsertion. Fixation of the graft with Kirschner wire and cortical screw is associated with lateral support plate and medial locked plate (**Figures 1–4**).

3.2.2 Cartilage regeneration techniques

3.2.2.1 Osteochondral allograft case 1

A 45-year-old male presented to us with 1 year of posteromedial right ankle pain. He reported pain related to physical activities, with a history of previous trauma, swelling of the joint, no feeling of instability, or joint blockage. Physical examination showed diffuse tenderness of the joint during maximal flexion and areas sensitive to touch in the medial tibiotalar joint line with negative ankle stability test (**Figures 5** and **6**).

The patient did not show good evolution with nonoperative treatment; a mosaicplasty with medial malleolus osteotomy was indicated (**Figures 7–11**).

The patient progressed well and returned the physical activities, including running without pain in the 6 month postoperatively.

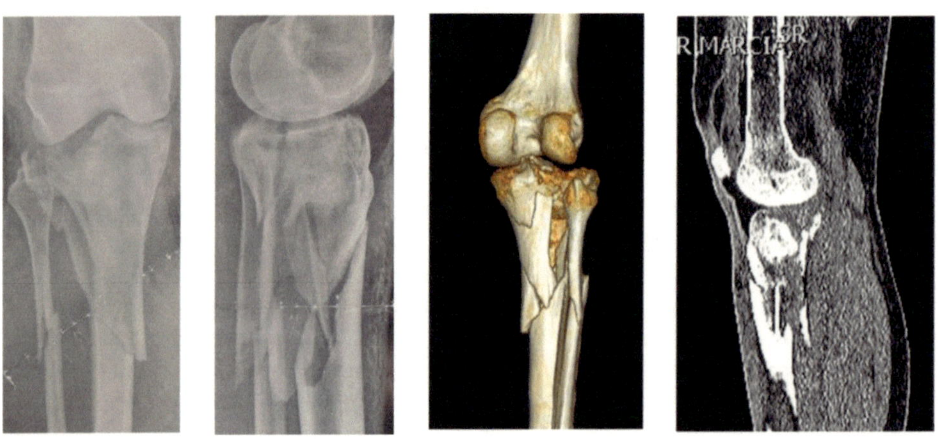

Figure 1.
Image arrival to the service.

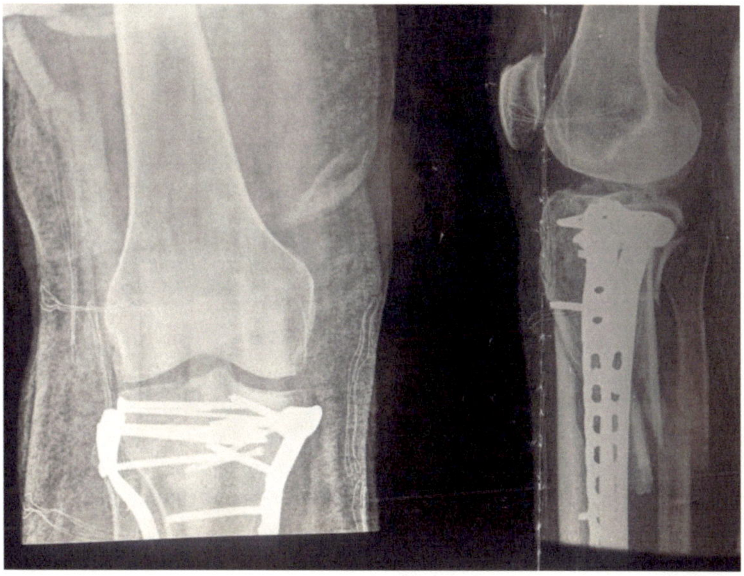

Figure 2.
Immediate postoperative.

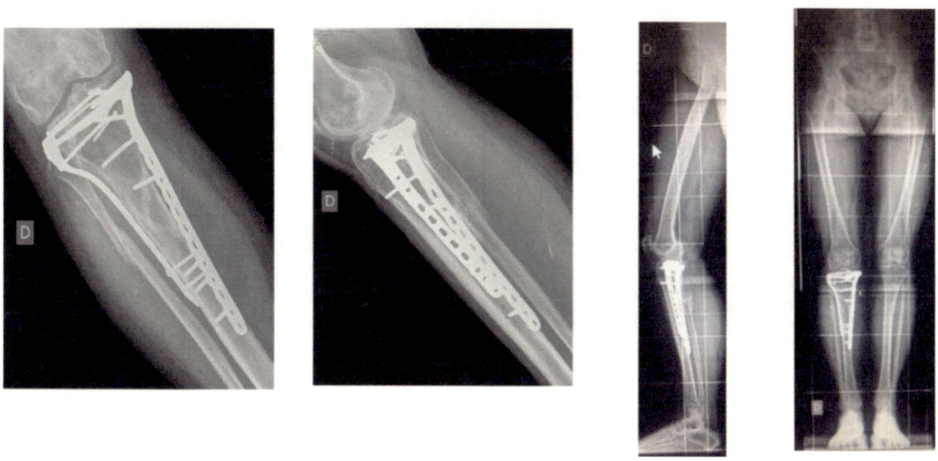

Figure 3.
An 18th month of evolution.

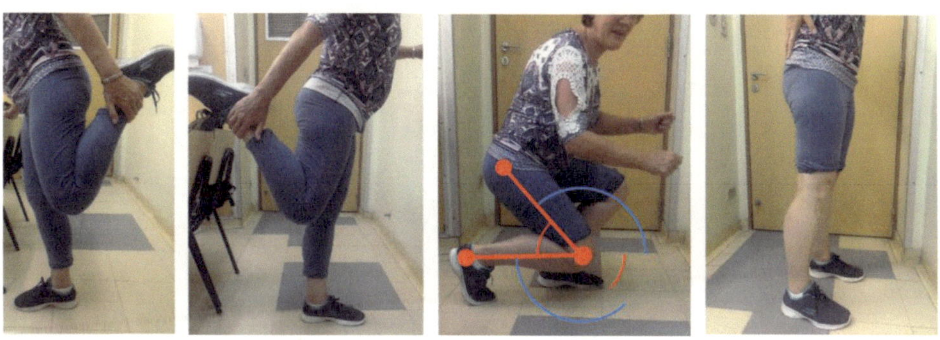

Figure 4.
Final result.

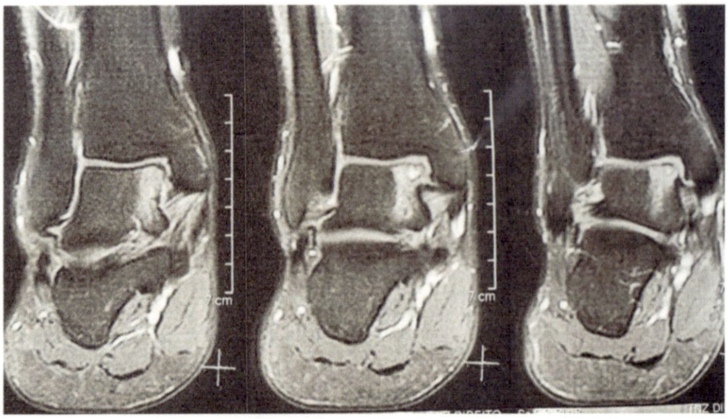

Figure 5.
Osteochondral lesion with cystic formation in the domus talar medialis measuring 1.0 × 0.7 × 0.7 cm, surrounded by area of bone edema. Stage V by Berndt & Harty.

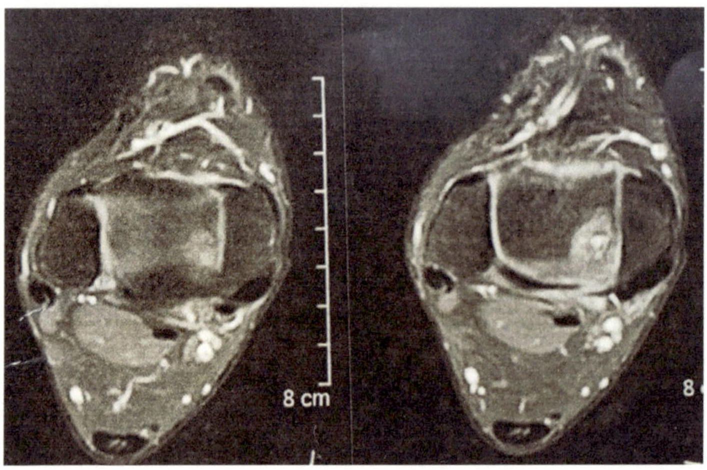

Figure 6.
Osteochondral lesion contained in zone 7 of Raikin.

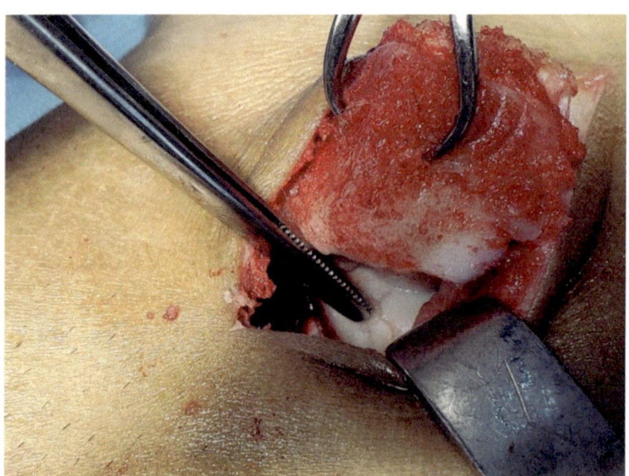

Figure 7.
A osteochondral lesion on the talar medial shoulder after osteotomy of the medial malleolus.

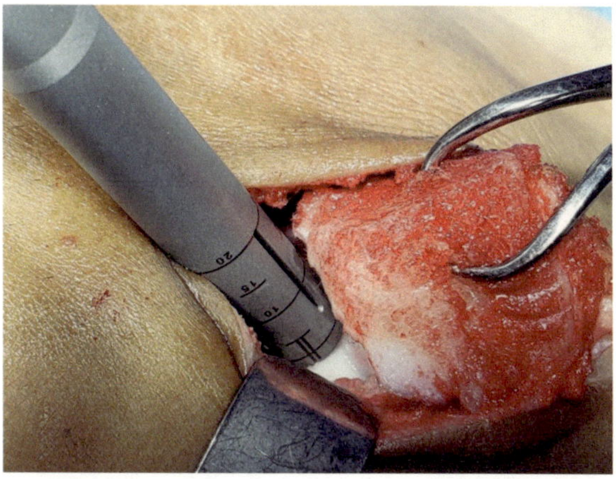

Figure 8.
Intraoperative image of cartilage defect removal.

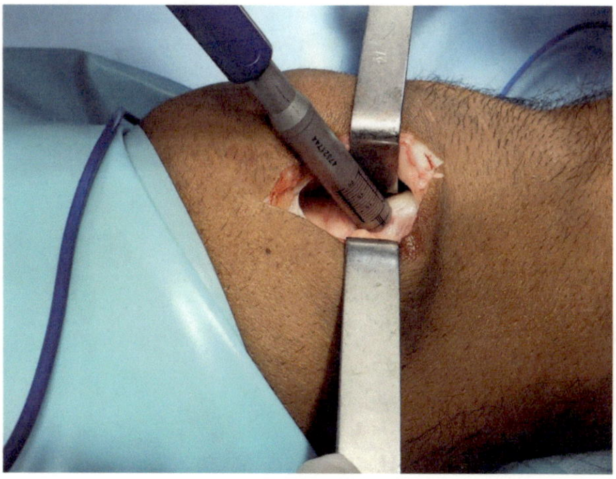

Figure 9.
Removal of the cylinder from the lateral superior region of the femoral trochlea (donor area).

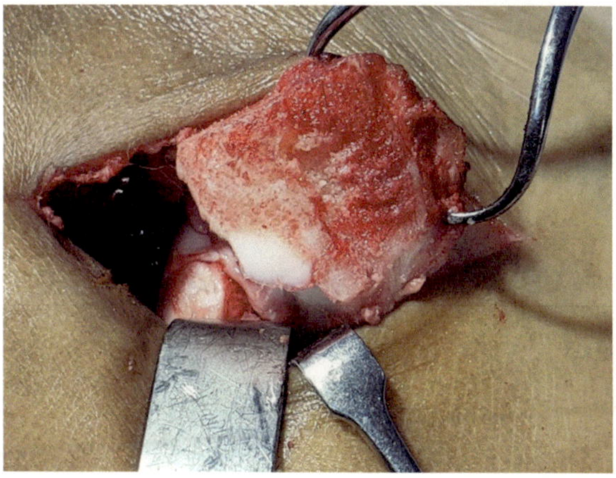

Figure 10.
Osteochondral cylinders inserted perpendicularly to the receiving surface.

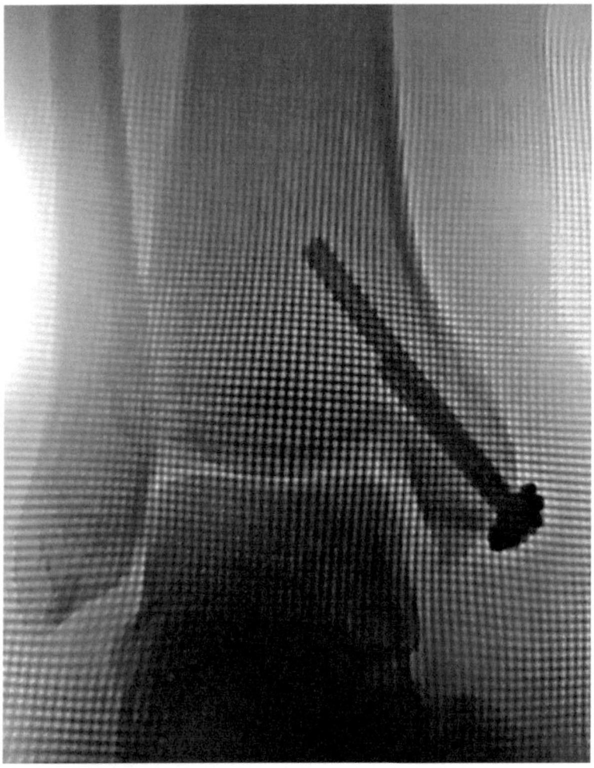

Figure 11.
Postoperative radiography.

4. Cartilage regeneration techniques

4.1 Autologous chondrocyte implantation (ACI)

The technique initially described by Brittberg in 1994 nowadays is the most used for cartilage regeneration [60]. ACI is a two-stage procedure; arthroscopy is initially performed to evaluate the lesion, and three to four CA chondral biopsies are taken from non-weight-bearing surfaces of the joint (intercondylar notch, peripheral edges of the femoral condyles). The sample is then sent to the laboratory, where chondrocytes are isolated with an enzymatic process. The chondrocytes are then cultured for 3–4 weeks until the volume increases by 30 times for implantation (12 million chondrocytes approx.). Usually, after 6 weeks of the initial surgery, the second procedure is performed [59–66].

4.1.1 First generation

Access with medial or lateral patellar arthrotomy is performed in association with defect debridement. A periosteal flap is removed from the proximal region of the tibia or medial femoral condyle. The flap is then attached to the defect (with its cambium layer facing the bone) on all sides, except at the top. The cultured chondrocytes are then injected under the flap, and, finally, the flap is then attached superiorly as well. The fibrin glue can be used to seal the edges of the flap [60].

4.1.2 Second generation

Due to complications arising from calcifications, the periosteum was replaced by a reabsorbable collagen membrane [66].

4.1.3 Third generation

The modification is the cultivation of the articular cells directly on a surface of a membrane-like MACI or cells grown within a scaffold [67].

4.1.4 Surgical technique

This procedure is a two-stage technique in which an arthroscopic approach is performed to evaluate the lesion and second used to harvest a sample of normal articular cartilage from a non-weight-bearing region of the knee. Chondrocytes are then isolated, cultured, and seeded onto a hyaluronan-based scaffold or collagen. The chondrocytes are then cultured for 3–4 weeks until the volume increases by 30 times for implantation, the second stage of the procedure arthrotomy to implant the scaffold in the lesion site. The chondral defect is prepared and is then used to shape the scaffold, which is pressed into the lesion site and secured with a thin layer of fibrin glue. The graft is assessed for stability before the wound is closed.

The best postoperative rehabilitation protocols are those of 6 weeks, starting the first 2 weeks with a partial load of 20% of body weight, followed by a progressive increase to a full load at 6 weeks postoperatively [67–73].

5. Conclusion

The articulations in their particularities refine the movement and enable a series of domains and skills of great importance for the development of human activities.

Thus, there is a growing interest in achieving more promising techniques in joint maintenance through cartilage repair. Some more modern techniques, involving the development and application of stem cells or the use of vectors to carry chondrocytes to the target lesion, still lack more consistent evidence in the long term that can justify their costs. Techniques that employ older, established concepts, such as microfracture and abrasion arthroplasties, are more accessible but fail to maintain their initial results over the years.

In this way, the development of a less invasive technique, aimed at preserving joint functions and minimizing symptoms, with sustainable durability and feasible cost, continues to guide the search for innovations in the arid terrain of joint preservation.

Conflict of interest

The author declares that have no competing interests.

Acronyms and abbreviations

OA	osteoarthritis
ACI	autologous chondrocyte implantation

Author details

Lourenço Galizia Heitzmann
Group of Orthopedics and Traumatology of the State Public Hospital of São Paulo, IAMSPE Institution(s), São Paulo, Brazil

*Address all correspondence to: lourenco@heitzmann.com.br

References

[1] Hootman JM, Helmick CG. Projections of US prevalence of arthritis and associated activity limitations. Arthritis and Rheumatism. 2006;**54**(1):226-229. DOI: 10.1002/art.21562

[2] Hunter W. Of the structure and disease of articulating cartilages. Philosophical Transactions of the Royal Society of London. 1743;**42B**:514-521

[3] Haasper C, Zeichen J, Meister R, Krettek C, Jagodzinski M. Tissue engineering of osteochondral constructs in vitro using bioreactors. Injury. 2008;**39**(1):66-76

[4] Hjelle K, Solheim E, Strand T, Muri R, Brittberg M. Articular cartilage defects in 1,000 knee arthroscopies. Arthroscopy. 2002;**18**(7):730-734

[5] Aroen A, Loken A, Heir S, et al. Articular cartilage lesions in 993 consecutive knee arthroscopies. The American Journal of Sports Medicine. 2004;**32**(1):211-215

[6] Shah MR, Kaplan KM, Meislin RJ, Bosco JA. Articular cartilage restoration of the knee. Bulletin of the NYU Hospital for Joint Diseases. 2007;**65**(1):51-60

[7] Guettler JH, Demetropooulos C, Yang K, et al. Osteochondral defects in the human knee: Influence of defect size on cartilage rim stress and load redistribution to surrounding cartilage. The American Journal of Sports Medicine. 2004;**32**:1451-1458

[8] Shelbourne KD, Jari S, Gray T. Outcome of untreated traumatic articular cartilage defects of the knee: A natural history study. The Journal of Bone and Joint Surgery. American Volume. 2003;**85**:8-16

[9] Mankin HJ. The response of articular cartilage to mechanical injury. The Journal of Bone and Joint Surgery. American Volume. 1982;**64**:460-466

[10] Camp CL, Stuart MJ, Krych AJ. Current concepts of articular cartilage restoration techniques in the knee. Sports Health. 2014;**6**:265-273

[11] Sahlström A, Johnell O, Redlund-Johnell I. The natural course of arthrosis of the knee. Clinical Orthopaedics and Related Research. 1997;**340**:152-157

[12] Huber M, Trattnig S, Lintner F. Anatomy, bio-chemistry, and physiology of articular cartilage. Investigative Radiology. 2000;**35**(10):573-580

[13] Zhou S, Cui Z, Urban JP. Factors influencing the oxygen concentration gradient from the synovial surface of articular cartilage to the cartilage-bone interface: A modeling study. Arthritis and Rheumatism. 2004;**50**(12):3915-3924

[14] Gardner OF, Archer CW, Alini M, Stoddart MJ. Chondrogenesis of mesenchymal stem cells for cartilage tissue engineering. Histology and Histopathology. 2013;**28**(1):23-42

[15] Leeb BF, Schweitzer H, Montag K, Smolen JS. A metaanalysis of chondroitin sulfate in the treatment of osteoarthritis. Journal of Rheumatology. 2000;**27**(1):205-211

[16] Sawitzke A, Shi H, Finco M, et al. Clinical efficacy and safety of glucosamine, chondroitin sulphate, their combination, celecoxib or placebo taken to treat osteoarthritis of the knee: 2-year results from GAIT. Annals of the Rheumatic Diseases. 2010;**69**:1459-1464

[17] Duivenvoorden T, Brouwer RW, van Raaij TM, Verhagen AP, Verhaar JAN, Bierma-Zeinstra SMA. Braces and orthoses for treating osteoarthritis of the knee. Cochrane Database of

Systematic Reviews. 2015;**3**:CD004020. DOI: 10.1002/14651858.CD004020.pub3

[18] Bartels E, Bliddal H, Schöndorff P, Altman R, Zhang W, Christensen R. Symptomatic efficacy and safety of diacerein in the treatment of osteoarthritis: A meta-analysis of randomized placebo-controlled trials. Osteoarthritis and Cartilage. 2010;**18**:289-296

[19] Altman RD, Bedi A, Karlsson J, Sancheti P, Schemitsch E. Product differences in intra-articular hyaluronic acids for osteoarthritis of the knee. The American Journal of Sports Medicine. 2016;**44**(8):2158-2165

[20] Steinmeyer J, Konttinen Y. Oral treatment options for degenerative joint disease-presence and future. Advanced Drug Delivery Reviews. 2006;**58**:168-211

[21] Richter DL, Schenck R, Wascher D, et al. Knee articular cartilage repair and restoration techniques: A review of the literature. Sports Health. 2016;**8**(2):153-160

[22] Insall J. The Pridie debridement operation for osteoarthritis of the knee. Clinical Orthopaedics and Related Research. 1974;(101):61-67

[23] Johnson LL. Arthroscopic abrasion arthroplasty historical and pathologic perspective: Present status. Arthroscopy. 1986;**2**:54-69. Available form: http://jbjs.org/public/instructionsauthors.aspx

[24] Ficat RP, Ficat C, Gedeon P, Toussaint JB. Spongialization: A new treatment for diseased patellae. Clinical Orthopaedics and Related Research. 1979;(144):74-83

[25] Frisbie DD, Trotter GW, Powers BE, Rodkey WG, Steadman JR, Howard RD, et al. Arthroscopic subchondral bone plate micro-fracture technique augments healing of large chondral defects in the radial carpal bone and medial femoral condyle of horses. Veterinary Surgery. 1999;**28**:242-255

[26] Burman MS, Finkelstein H, Mayer L. Arthroscopy of the knee joint. Journal of Bone and Joint Surgery. 1934;**16A**:255-268

[27] Magnuson PB. Joint debridement. Surgical treatment of degenerative arthritis. Surgery, Gynecology & Obstetrics. 1941;**73**:1-9

[28] Pridie KH. A method of resurfacing osteoarthritic knee joints. The Journal of Bone and Joint Surgery. American Volume. 1959;**41-B**:618-619

[29] Johnson LL. Diagnostic and Surgical Arthroscopy. MO, USA: CV Mosby; 1980

[30] Steadman RJ, Rodkey WG, Singleton SB, Briggs KK. Microfracture technique for full thickness chondral defects: Technique and clinical results. Operative Techniques in Orthopaedics. 1997;7:300-305

[31] Akizuki S, Yasukawa Y, Takizawa T. Does arthroscopic abrasion arthroplasty promote cartilage regeneration in osteoarthritic knees with eburnation? A prospective study of high tibial osteotomy with abrasion arthroplasty versus high tibial osteotomy alone. Arthroscopy. 1997;**13**:9-17

[32] Steadman JR, Rodkey WG, Briggs KK, Rodrigo JJ. The microfracture technic in the management of complete cartilage defects in the knee joint. Der Orthopade. 1999;**28**(1):26-32, German

[33] Suh JK, Scherping S, Marui T, Steadman JR, Woo S-LY. Basic science of articular cartilage injury and repair. In: Sledge SL, editor. Operative Techniques in Sports Medicine. Management of Osteochondral Injuries. Philadelphia: WB Saunders Company; 1995. pp. 78-86

[34] Salisbury RB, Mc Mahon MR. Joint debridement and abrasion arthroplasty for degenerative joint disease of the knee joint. In: Parisien SJ, editor. Current Techniques in Arthroscopy. Philadelphia: Current Medicine; 1994. pp. 147-167

[35] Steadman JR, Rodkey WG, Briggs KK. Microfracture to treat full-thickness chondral defects: Surgical technique, rehabilitation, and outcomes. The Journal of Knee Surgery. 2002;**15**:170-176

[36] Kreuz PC, Erggelet C, Steinwachs MR, Krause SJ, Lahm A, Niemeyer P, et al. Is microfracture of chondral defects in the knee associated with different results in patients aged 40 years or younger? Arthroscopy. 2006;**22**(11):1180-1186

[37] Mithoefer K, McAdams T, Williams RJ, Kreuz PC, Mandelbaum BR. Clinical efficacy of the microfracture technique for articular cartilage repair in the knee: An evidence-based systematic analysis. The American Journal of Sports Medicine. 2009;**37**(10):2053-2063

[38] Kreuz PC, Steinwachs MR, Erggelet C, Krause SJ, Konrad G, Uhl M, et al. Results after microfracture of full-thickness chondral defects in different compartments in the knee. Osteoarthritis and Cartilage. 2006;**14**(11):1119-1125

[39] Frehner F, Benthien JP. Microfracture: State of the art in cartilage surgery? Cartilage. 2017. DOI: 10.1177/1947603517700956

[40] Pap K, Krompecher I. Arthroplasty of the knee—Experimental and clinical experiences. Journal of Bone and Joint Surgery. 1961;**43A**:523-537

[41] Wagner H. Operative behandlung der osteochondrosis dissecans der kniegelenkes. Zeitschrift für Orthopädie. 1964;**98**:333-355

[42] Muller W. Osteochondritis dissecans. In: Hastings DW, editor. Progress in Orthopaedics Surgery. New York: Springer; 1978. p. 135

[43] Matsusue Y, Yamamuro T, Hiromichi H. Case report. Arthroscopic multiple osteochondral transplantation to the chondral defect in the knee associated with anterior cruciate ligament disruption. Arthroscopy. 1993;**9**:318-321

[44] Hangody L, Kish G, Kárpáti Z, Szerb I, Udvarhelyi I. Arthroscopic autogenous osteochondral mosaicplasty for the treatment of femoral condylar articular defects. A preliminary report. The Knee Surgery, Sports Traumatology, Arthroscopy. 1997;**5**(4):262-267

[45] Hangody L, Kish G, Karpati Z, Udvarhelyi I, Szigeti I, Bely M. Mosaicplasty for the treatment of articular cartilage defects: Application in clinical practice. Orthopedics. 1998;**21**:751-756

[46] Gudas R, Kalesinskas R, Kimtys V, et al. A prospective randomized clinical study of mosaic osteochondral autologous transplantation versus microfracture for the treatment of osteochondral defects in the knee joint in young athletes. Arthroscopy. 2005;**21**:1066-1075

[47] Hangody L, Vásárhelyi G, Hangody LR, Sükösd Z, Tibay G, Bartha L, et al. Autologous osteochondral grafting—Technique and long-term results. Injury. 2008;**9**(1):32-39. DOI: 10.1016/j.injury.2008.01.041

[48] Morelli M, Nagamori J, Miniaci A. Management of chondral injuries of the knee by osteochondral autogenous transfer (mosaicplasty). The Journal of Knee Surgery. 2002;**15**:185-190

[49] Ozturk A, Ozdemir MR, Ozkan Y. Osteochondral autografting (mosaicplasty) in grade IV cartilage defects in the knee joint: 2- to 7-year

results. International Orthopaedics. 2006;**30**(3):200-204

[50] Bartz RL, Kamaric E, Noble PC, Lintner D, Bocell J. Topographic matching of selected donor and recipient sites for osteochondral autografting of the articular surface of the femoral condyles. The American Journal of Sports Medicine. 2001;**29**:207-212

[51] Jerosch J, Filler T, Peuker E. Is there an option for harvesting autologous osteochondral grafts without damaging weight-bearing areas in the knee joint? Knee Surgery, Sports Traumatology, Arthroscopy. 2001;**8**:237-240

[52] Menche DS, Vangsness CT Jr, Pitman M, Gross AE, Peterson L. The treatment of isolated articular cartilage lesions in the young individual. Instructional Course Lectures. 1998;**47**:505-515

[53] Marcacci M, Kon E, Zaffagnini S, et al. Multiple osteochondral arthroscopic grafting (mosaicplasty) for cartilage defects of the knee: Prospective study results at 2-year follow-up. Arthroscopy. 2005;**21**(4):462-470

[54] Reddy S, Pedowitz DI, Parekh SG, Sennett BJ, Okereke E. The morbidity associated with osteochondral harvest from asymptomatic knees for the treatment of osteochondral lesions of the talus. The American Journal of Sports Medicine. 2007;**35**(1):80-85

[55] Bentley G, Biant LC, Carrington RW. A prospective, randomised comparison of autologous chondrocyte implantation versus mosaicplasty for osteochondral defects in the knee. Journal of Bone and Joint Surgery. 2003;**85B**:223-230

[56] O'Driscoll SW. The healing and regeneration of articular cartilage. The Journal of Bone and Joint Surgery. American Volume. 1998;**80**:1795-1812

[57] O'Driscoll SW. Articular cartilage regeneration using periosteum. Clinical Orthopaedics. 1999;**367**:S186-S203

[58] Fonseca F. Mosaicoplastia revestida com periósseo no tratamento de perda osteocondral do joelho [Mosaicplasty with periosteal graft for resurfacing local full-thickness chondral defects of the knee]. Revista Brasileira de Ortopedia. 2009;**44**(2):153-158. Available from: http://www.scielo.br/scielo.php?script=sci%7B_%7Darttext%7B&%7Dpid=S0102-36162009000200011%7B&%7Dlang=pt

[59] Brittberg M, Nilsson A, Lindahl A, Ohlsson C, Peterson L. Rabbit articular cartilage defects treated with autologous cultured chondrocytes. Clinical Orthopaedics and Related Research. 1996;**326**:270-283

[60] Brittberg M, Lindahl A, Nilsson A, Ohlsson C, Isaksson O, Peterson L. Treatment of deep cartilage defects in the knee with autologous chondrocyte transplantation. The New England Journal of Medicine. 1994;**331**(14):889-895

[61] Brooks PJ. Role of osteochondral transplantation in the treatment of joint surface defects. Sports Medicine. 1994

[62] Niemeyer P, Andereya S, Angele P, Ateschrang A, Aurich M, Baumann M, et al. Autologous chondrocyte implantation (ACI) for cartilage defects of the knee: A guideline by the working group "tissue regeneration" of the German Society of orthopaedic surgery and traumatology (DGOU). Zeitschrift für Orthopädie und Unfallchirurgie. 2013;**151**(1):38-47

[63] Valderrabano V, Miska M, Leumann A, Wiewiorski M. Reconstruction of osteochondral lesions of the talus with autologous spongiosa grafts and autologous matrix-induced chondrogenesis. The American Journal of Sports Medicine. 2013;**41**(3):519-527

[64] Saris DB, Vanlauwe J, Victor J, Almqvist KF, Verdonk R, Bellemans J, et al. Characterized chondrocyte implantation results in better structural repair when treating symptomatic cartilage defects of the knee in a randomized controlled trial versus microfracture. The American Journal of Sports Medicine. 2007;**36**:235-246

[65] Bentley G, Biant LC, Vijayan S, Macmull S, Skinner JA, Carrington RW. Minimum ten-year results of a prospective randomised study of autologous chondrocyte implantation versus mosaicplasty for symptomatic articular cartilage lesions of the knee. Journal of Bone and Joint Surgery. British Volume (London). 2012;**94**(4):504-509

[66] Macmull S, Jaiswal PK, Bentley G, Skinner JA, Carrington RW, Briggs TW. The role of autologous chondrocyte implantation in the treatment of symptomatic chondromalacia patellae. International Orthopaedics. 2012;**36**:1371-1377

[67] Welton KL, Logterman S, Bartley JH, Vidal AF, McCarty EC. Knee cartilage repair and restoration: Common problems and solutions. Clinics in Sports Medicine. 2018;**37**(2):307-330

[68] Ebert JR, Edwards PK, Fallon M, Ackland TR, Janes GC, Wood DJ. Two-year outcomes of a randomized trial investigating a 6-week return to full weightbearing after matrix-induced autologous chondrocyte implantation. The American Journal of Sports Medicine. 2017;**45**(4):838-848

[69] Ebert JR, Fallon M, Robertson W, et al. Radiological assessment of accelerated versus traditional approaches to postoperative rehabilitation following matrix-induced autologous chondrocyte implantation. Cartilage. 2011;**2**(1):60-72

[70] Ebert JR, Fallon M, Zheng MH, Wood DJ, Ackland TR. A randomized trial comparing accelerated and traditional approaches to postoperative weightbearing rehabilitation after matrix-induced autologous chondrocyte implantation: Findings at 5 years. The American Journal of Sports Medicine. 2012;**40**(7):1527-1537

[71] Ebert JR, Robertson WB, Lloyd DG, Zheng MH, Wood DJ, Ackland T. A prospective, randomized comparison of traditional and accelerated approaches to postoperative rehabilitation following autologous chondrocyte implantation: 2-year clinical outcomes. Cartilage. 2010;**1**(3):180-187

[72] Edwards PK, Ackland T, Ebert JR. Accelerated weightbearing rehabilitation after matrix-induced autologous chondrocyte implantation in the tibiofemoral joint: Early clinical and radiological outcomes. The American Journal of Sports Medicine. 2013;**41**(10):2314-2324

[73] Kraeutler MJ, Belk JW, Carver TJ, McCarty EC. Is delayed weightbearing after matrix-associated autologous chondrocyte implantation in the knee associated with better outcomes? A systematic review of randomized controlled trials. Orthopaedic Journal of Sports Medicine. 2018;**6**(5):1-10

Chapter 6

Current Tissue Engineering Approaches for Cartilage Regeneration

He Huang, Hongyao Xu and Jianying Zhang

Abstract

Cartilage is one of the critical tissues existed in human and animal bodies. Unlike most tissues, cartilage does not have blood vessels, nerves, and lymphatics. Most cartilage tissues *in vivo* are subjected to large mechanical loads, and its principal function is to provide a smooth and lubricated surface to facilitate the transmission of mechanical loads with a low frictional coefficient. As a result, cartilage tissues are easily injured. Cartilage defects are frequently caused by trauma, aging, congenital diseases (osteochondritis), and many more factors such as endocrine pathologies and cancer. The damaged cartilage has a limited capacity for healing and repairing. Thus, restoration of normal structure and function to damaged cartilage is one of the most challenging areas in orthopedic research and sports medicine. Tissue engineering provides a prospective alternative strategy by seeding chondrogenic cells into or onto biocompatible scaffolds to produce engineer cartilage for damaged cartilage repair. This book chapter has summarized recent progress in cartilage tissue engineering including stem cells, growth factors, bioactive molecules, and biomaterial scaffolds used for cartilage regeneration. The procedures for some new approaches have also been described.

Keywords: chondrogenesis, cartilage tissue engineering, stem cells, growth factors, platelet-rich plasma, bioactive molecules, biomaterial scaffold

1. Introduction

Cartilage is one of the critical tissues existed in human and animal bodies, such as rib cage, ear, nose, bronchial tubes, intervertebral discs, meniscus, and the joints between bones [1]. Cartilage injuries are the most common diseases. According to National Health Interview Survey (NHIS), in 2010–2012, about 52.5 million adults in the USA had doctor-diagnosed arthritis, and by 2040, the number of US adults with doctor-diagnosed arthritis is projected to increase 49% to 78.4 million. That means about 25.9% of all adults have arthritis [2, 3]. Degeneration of the intervertebral disc, a fibrocartilaginous joint residing between adjacent vertebrae in the vertebral column, is the most frequent cause of low back pain and another significant cartilage-related disease [4]. The overall cost of chronic low back pain exceeds the combined costs of stroke, respiratory infection, diabetes, coronary artery disease, and rheumatoid disease [5]. However, the damaged cartilage has little ability for repairing itself due to the lack of blood supply, nerves, and lymphangion [1], and the effective therapeutic treatments for cartilage regeneration are very few.

Tissue engineering is an interdisciplinary field that applies the principles of engineering and the life sciences toward the development of biological substitutes that restore, maintain, or improve tissue function [6]. Stem cells, scaffold, and biologic active molecules are three key components in tissue engineering [7]. Successful tissue engineering relies on multiple factors including obtaining appropriate cells for implantation, directing the development of those cells on an appropriate differentiated pathway using growth factors and/or cytokines, supporting the growing cells on a three-dimensional matrix, and having that matrix remains in the injured tissue area, at least until healing is completed [6, 8]. This book chapter highlights the recent developments of tissue engineering approaches including stem cells, biomaterials, bioactive compounds, and reagents used for cartilage regeneration and repair.

2. Stem cells used for cartilage regeneration

Stem cells have multidifferentiation potential, which can differentiate into distinctive end-stage cell types including bone, cartilage, muscle, bone marrow stroma, tendon/ligament, fat, dermis, and other connective tissues [9]. There are many cell types that have been manipulated *in vitro* and subsequently implanted to repopulate a cartilage defect. It must be ensured that the implanted cells are immunoprivileged or provide immunosuppressive agents to avoid rejection by the host immune system.

2.1 Autologous chondrocytes

Autologous chondrocytes were first used for the treatment of cartilage defects of the patients by a Swedish group in 1994 [10]. This approach needs a slice of healthy articular cartilage obtained arthroscopically from proximal part of the medical femoral condyle of the affected knee joint during the first operation [11]. The chondrocytes were isolated from this healthy articular cartilage and cultured for 2–3 weeks to prepare enough cells (about 5×10^6) for damaged cartilage repair [11]. The clinical studies have shown that the treatment of autologous chondrocytes prompts pain reduction, improves quality of life, and delays the need of joint replacement in many cases [12–14]. Despite the encouraging clinical results, there are still limitations to the use of autologous chondrocyte transplantation. The conventional technique is accompanied with periosteum harvest and fixation over the cartilage defects via large skin incisions. Autologous chondrocytes were injected underneath the periosteal flap. Hypertrophy of the periosteum with high rate of revision arthroscopies and the risk of transplant failure of up to 20% are major drawbacks of the conventional autologous chondrocyte transplantation [14]. Moreover, the complexity and the cost of the two surgical procedures, the biological response of the periosteal flap, and the de-differentiation and consequent capacity loss associated with *in vitro* expansion of isolated chondrocytes are also the limitations [15].

2.2 Bone marrow-derived mesenchymal stem cells (BMSCs)

Mesenchymal stem cells (MSCs) are multipotent stromal cells first identified and described in 1966 by Alexander Fridenstein [16, 17]. Adult MSCs were originally isolated from bone marrow in 1999 by Pittenger and his colleagues [18]. Subsequent studies have demonstrated that MSCs present in various parts of the body including bone marrow (BM), peripheral blood, umbilical cord blood, fatty tissues, skeletal and cardiac muscles, Wharton's Jelly of umbilical cord, facet joints, interspinous ligaments, and ligamentum flavum [19–23]. Many studies have shown that MSCs can migrate to injury sites, induce peripheral tolerance, and inhibit the release of proinflammatory cytokines. It has been demonstrated that MSCs can also promote tissue repair and survival of damaged cells [24]. However, it is not clear which adult tissue-derived MSCs should be used as a good source for cartilage repair.

Autologous bone marrow mesenchymal stem cell (BMSCs) transplantation was first used for the repair of full-thickness articular cartilage defects in human patellae by a Japanese group [25]. BMSCs were aspirated from iliac crest and the nucleated cells were cultured. Adherent cells were subsequently collected, embedded in a collagen gel, transplanted into the articular cartilage defect in patellae, and covered with autologous periosteum. Six months after transplantation, clinical symptoms (pain and walking disability) were improved and the improvement was persisted for 9 years post-transplantation [26]. Sixteen years after transplantation, no clinical problem has been reported. Human autologous BMSCs have been used successfully to treat articular cartilage defects. Twelve months after BMSC transplantation, magnetic resonance imaging (MRI) revealed complete defect fill and complete surface congruity with native cartilage [27]. Currently, autologous BMSC transplantation has been widely used for cartilage repair [26, 28, 29]. Although BMSC treatment did not require any cell expansion or manipulation, reducing costs, and risks involved, the quantity of bone marrow cells was somewhat unsatisfactory [16].

2.3 Adipose-derived stem cells (ADSCs)

Among MSCs, adipose-derived stem cells (ADSCs) have been recognized as an appropriate cell type with chondrogenic potential and high proliferation capacity [30, 31]. Approximately 400,000 liposuction surgeries are performed in the USA each year, and these procedures yield anywhere from 100 ml to 3 liters of lipoaspirate tissue [32]. This material is routinely discarded. It is well known that adipocytes are developed from mesenchymal cells via a complex cascade of transcriptional and non-transcriptional events that occur throughout the human life. Thus, adipose tissue is a good stem cell source.

The initial methods to isolate cells from adipose tissue were developed by Rodbell and colleagues [33]. They isolated adipose-derived stromal cells from rat fat pads by four steps. Step 1: Adipose tissue was minced into small pieces. Step 2: The adipose tissue pieces were digested with collagenase. Step 3: The cell pellet was obtained by centrifuge. Step 4: The cell pellet was cultured for future use. This protocol has been widely used for the isolation of adipose-derived stem cells (ADSCs) from human adipose tissues with some modifications [34, 35].

The adipose tissue can be collected by needle biopsy or liposuction aspiration. The collected adipose tissues should be washed with 5% penicillin/streptomycin (P/S)-containing phosphate-buffered saline (PBS) twice, and then the tissue samples should be put in a sterile tissue culture plate and cut into small pieces. The minced tissues are digested with 0.075% collagenase at 37°C for 30 min; the collagenase is removed by centrifuging the digested solution (adipose tissue and collagenase mixture) at 1200 g for 10 min; the adipose-derived stem cells-containing pellet is then resuspended with culture medium (alpha-MEM, Mediatech, Herndon, VA) supplemented with 20% of fetal bovine serum (FBS), 1% L-glutamine (Mediatech, Herndon, VA), and 1% penicillin/streptomycin (Mediatech, Herndon, VA). The cell suspension is filtered through 70-μm cell strainer and cultured in a humidified tissue culture incubator at 37°C with 5% CO_2. The medium is changed every second day until the cells reach 80–90% confluence. It is important that the adipose tissue should be treated within 24 hours, and the cells isolated from about 500 mg of adipose tissue should be added into one well of 12-well plates.

Adipose-derived stem cells (ADSCs) are readily accessible with no morbidity and display the capability to differentiate into several cell lineages, including the spontaneous chondrogenic differentiation [30]. Compared with bone marrow-derived MSCs, adipose-derived MSCs from lipoaspirates are acquired using a less invasive procedure and are in large amounts [36]. ADSCs have been used for the repair of articular cartilage defect in nonweight-bearing areas [37].

2.4 Synovial-derived stem cells (SDSCs)

Synovial-derived MSCs have been isolated from human synovial fluid and synovium of the knee and the hip using the following protocols [38, 39]. The synovial tissue samples (wet weight 10–50 mg) were obtained aseptically from the joints and rinsed twice with Hanks' balanced salt solution (HBSS; Life Technologies, Carlsbad, CA) supplemented with antibiotic-antimycotic solution (100 units/ml penicillin, 100 μg/ml streptomycin, and 0.25 μg/ml amphotericin B; Life Technologies; Carlsbad, CA). The washed tissues were minced into small pieces and digested with 0.5 ml of 0.2% collagenase (Life Technologies, Carlsbad, CA) in high-glucose Dulbecco's Modified Eagle's Medium (DMEM; Life Technologies, Carlsbad, CA) at 37°C for 1 hour. The digested solution were removed by centrifugation at 1500 g for 10 min; the SDSCs-containing pellet was resuspended in growth medium (high-glucose DMEM supplemented with 10% FBS and 100 units/ml penicillin, 100 μg/ml streptomycin) and cultured in a humidified tissue culture incubator at 37°C with 5% CO_2. The medium was first changed at day 7 and changed every 3 days until the cells reach 80–90% confluence. It is important that the synovial tissue should be treated within 24 hours.

SDSCs obtained by above procedures have a higher proliferative capacity and chondrogenic potential than the MSCs derived from other sources [39, 40]. A small synovial tissue biopsy is an easily accessible source of autologous MSCs in the context of an explorative or therapeutic arthroscopy. These cells can be subsequently used for the regeneration of damaged cartilage. Autologous chondrocyte transplantation used for cartilage defect repair is limited by the availability of cells, particularly in elderly individuals, and by the well-known dedifferentiation events associated with chondrocyte expansion [39, 41]. Furthermore, SDSCs can be harvested relatively in a minimally invasive manner from synovial fluid and retain a particularly high capacity for chondrogenic differentiation and proliferation compared with MSCs obtained from other tissues, such as bone marrow or cartilage, those have second injury on healthy tissues. SDSCs may be an optimal alternative source of chondrogenic cells for cartilage defect repair.

A recent research has shown that xenogenic implantation of equine SDSCs into rat cartilage defect area leads to articular cartilage regeneration [42]. Horse joints are anatomically equivalent to the human knee and ankle; as a result, horses are widely used as large animal preclinical models for cartilage repair studies. However, large animal studies pose logistical and financial challenges, and small animal rodent models are cost-effective and have proven to be useful for proof-of-concept studies. There was no any immune response to the equine cells in the treated rat knees [42]. This result was also confirmed by a xenogenic transplantation of human MSCs in a critical size defect of the sheep tibia for bone regeneration [43]. Another xenogenic transplantation study has shown that human MSCs can enhance damaged pig intervertebral disc regeneration [44].

3. Growth factors used for cartilage regeneration

Growth factors play an important role in cartilage regeneration. Although some growth factors used in cartilage repair have been well documented [45], it is necessary to summarize the most important chondrogenic differentiation-related growth factors in this chapter.

3.1 Transforming growth factor-beta family (TGF-β)

In cartilage repair, the four most thoroughly investigated members of TGF-β superfamily are TGF-β1, TGF-β3, bone morphogenetic protein-2 (BMP-2), and bone morphogenetic protein-7 (BMP-7) [45, 46]. It has been reported that TGF-beta 1 stimulates chondrocyte synthetic activity and decreases the catabolic activity of IL-1 [47]. TGF-β3 has been used to simulate extracellular matrix (ECM) synthesis in rabbit cartilage injury [48]. Bone morphogenetic proteins (BMPs) play an important role in the development of bone and cartilage. They are involved in the hedgehog pathway, TGF beta signaling pathway, and in cytokine-cytokine receptor interaction. Animal studies have shown that BMP-2 enhanced cartilage matrix production and blocked the IL-1-induced cartilage degeneration [49].

BMP-7 is another gold standard growth factor for cartilage repair [50]. It has been reported that BMP-7 inhibits cell proliferation but stimulates ECM synthesis in both SDSCs and BMSCs [51, 52].

3.2 Insulin-like growth factor-I (IGF-I)

IGF-1 is a multifunctional growth factor. The studies have found that IGF-1 play an important role in maintaining articular cartilage integrity. IGF-I deficiency has led to the development of articular lesions [53]. IGF-1 can not only enhance the synthesis of proteoglycans and upregulate the gene expression of collagen II but also can reduce the degradation of extracellular matrixes by decreasing the production of matrix metallopetidase-13 (MMP-13) [54–56]. The research has shown that IGF-1 exerts these functions in a dose-dependent manner [57]. Low dose of IGF-1 has a beneficial effect on bone remodeling by increasing bone formation markers in serum [58]. Higher IGF-1 levels in osteoarthritis (OA) osteoblasts could be correlated with bone sclerosis [59].

3.3 Fibroblast growth factor (FGF)

There are two FGF members used in cartilage regeneration. One of them is called as basic FGF (bFGF) or FGF-2, and the other one is called as FGF-18. FGF-2 increases anabolic material levels and decreases aggrecanase levels in cartilage. In vivo study has indicated that bFGF can promote cartilage repair [60]. However, some study indicated that the concentration of FGF-2 in synovial fluid samples of OA patients is approximately twice that of normal healthy knee joints [61]. Further studies found that FGF-2 promoted the repair of partial thickness defects of articular cartilage in immature rabbits but not in mature rabbits [62].

A rat model study has shown that FGF-18 stimulates chondrogenesis and cartilage repair in a concentration-dependent manner [63]. More studies have demonstrated that FGF-18 may present a therapeutic agent for osteoarthritis [64, 65]. A recombinant form of human FGF-18 has been used for cartilage injury treatment [66].

3.4 Platelet-rich plasma (PRP)

Platelets play a fundamental role in hemostasis and are a natural source of growth factors. More than 30 growth factors have been identified in PRP; among them, the following six growth factors play an important role in cartilage

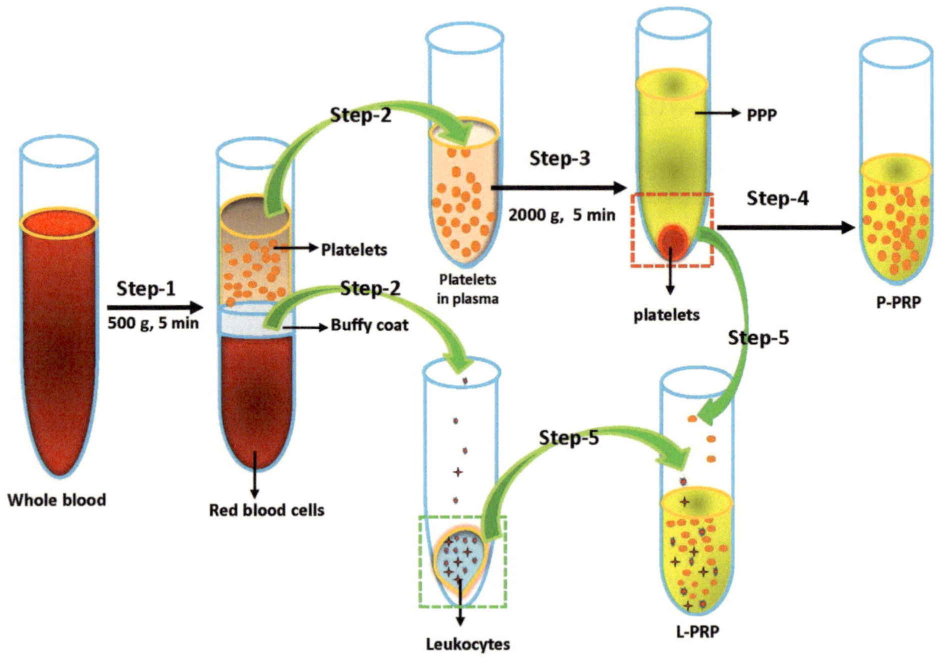

Figure 1.
Scheme of preparation of P-PRP and L-PRP from whole blood using five steps.

regeneration. They are TGF-β1, platelet-derived growth factor (PDGF), fibroblast growth factor (FGF), insulin-like growth factor 1 (IGF-1), epidermal growth factor (EGF), and vascular endothelial growth factor (VEGF) [67, 68].

The concentration of platelet in PRP used for cartilage repair should be two to three times higher than that of baseline [69]. PRP can be prepared by the following five procedures (**Figure 1**). **Step 1:** blood (9 parts) is added into 3.8% sodium citrate solution (1 part) in a centrifuge tube and centrifuged at 500 g for 5 min to obtain three layers. **Step 2:** The supernatant at the top layer is transferred into a new tube, which is called as platelets-containing plasma, and the middle layer is transferred into another new tube, which is called leukocytes-containing plasma. **Step 3:** The platelets-containing plasma is centrifuged at 2000 g for 5 min to separate platelet-poor plasma (PPP) from the platelet pellet. **Step 4:** The platelet pellet is resuspended with appropriate amount of PPP to make pure PRP (P-PRP). **Step 5:** The leukocytes-containing plasma is mixed with platelet pellet and resuspended with appropriate amount of PPP to make leukocytes-containing PRP (L-PRP). Both P-PRP and L-PRP can be used for cartilage tissue engineering [70].

4. Bioactive molecules used for cartilage tissue engineering

Bioactive molecules used in cartilage tissue engineering include two kinds of materials: one is small molecular weight bioactive compound and the other one is high molecular weight materials including some nature biomaterials and synthetic polymers. Both of them play critical role in cartilage tissue engineering.

4.1 Kartogenin (KGN)

Kartogenin (KGN), a small heterocyclic molecule, has been discovered to enhance chondrogenic differentiation of human MSCs by regulating the

CBFbeta-RUNX1 transcriptional program [71, 72]. Animal studies have shown that KGN can promote rabbit meniscus regeneration [73] and wounded rat enthesis repair [70, 74]. *In vitro* and *ex vivo* experiments showed that KGN can reduce nucleus pulposus cell degeneration induced by interleukin-1beta (IL-1β) and tumor necrosis factor-alpha [75]. More recent studies indicated that KGN inhibited pain behavior, chondrocyte inflammation, and attenuated osteoarthritis progression in mice [76]; enhanced collagen organization and mechanical strength of the repaired enthesis of mouse rotator cuff [77]; and induced chondrogenic differentiation of dental pulp stem cells [78].

These findings invigorate research into small-molecule therapy and regenerative medicine for cartilage diseases. It also provides new insights into the control of chondrogenesis that may ultimately lead to a stem cell-based therapy for osteoarthritis (OA). KGN and other structurally related small molecules that can promote selective differentiation of MSCs into chondrocytes may prove to be extremely useful for improving the outcome of cell-based therapy by stimulating endogenous mechanisms for repair of damaged cartilage, thus enhancing the joint's intrinsic capacity for cartilage regeneration [79].

4.2 Simvastatin

Simvastatin is a kind of HMG-CoA reductase inhibitor, which is widely used therapeutically to reduce morbidity and mortality in patients with hyperlipidemic cardiovascular disease [80]. In addition to lowing low-density lipoprotein (LDL) cholesterol, statins have broad-range pleiotropic effects, including anti-inflammatory effects, which could exert an effect on synovium and cartilage [81]. Animal studies found that simvastatin markedly inhibited not only developing but also established collagen-induced arthritis [82]. Simvastatin inhibited the IL-6 and TNF-α production of human chondrocytes and cartilage explants in a concentration-dependent manner. Higher concentrations of simvastatin decreased nitric oxide (NO) production in both of human chondrocytes and cartilage explants [83]. Statin treatment has also been shown to positively regulate components of the extracellular matrix in a rabbit OA model [84]. More studies have shown that local application of simvastatin enhanced tendon-bone interface healing in rabbits [85]. These studies have shown that the effect of simvastatin on articular chondrocytes may provide novel insight regarding the role of cholesterol homeostasis and signaling during cartilage development.

4.3 Biomaterial scaffolds for cartilage tissue engineering

Biomaterial scaffolds play an important role in cartilage tissue engineering, which act as a carrier to deliver the cells and bioactive molecules to the damaged tissue areas and also work as a template for tissue regeneration, to guide the growth of new tissue.

There are two groups of biomaterial scaffolds used for cartilage tissue engineering. They are synthetic polymers and natural polymers. Commonly used natural materials in cartilage repair are agarose, alginate, chitosan, collagen, fibrin, and hyaluronan.

Agarose is a galactose polymer, which is suitable for cell encapsulation, especially for chondrocytes. When the ADSCs were cultured in agarose, they were differentiated into chondrocytes as evidenced by upregulation of the production of glycosaminoglycan (GAG) [86]. Moreover, dynamically loaded cell-seeded agarose hydrogel provided better graft tissues in a repair model of full thickness defects in rabbit joint cartilage [87]. PRP combined with agarose as a bioactive scaffold has shown to enhance cartilage repair [88].

Another extensively studied natural scaffold used for cartilage tissue engineering is alginate, which is a polysaccharide extracted from brown algae. Generally, alginate is hydrophilic and water-soluble, thickening in neutral conditions, which is of great importance for *in situ* hydrogel formation [89]. The good gelling properties of alginate-based scaffolds allowed them to be used as an injectable scaffold for the damaged cartilage repair. Human dental pulp stem cells were cultured in 3% alginate hydrogel and implanted in a rabbit damaged cartilage area. Three months after surgery, significant cartilage regeneration was observed [90]. More studies have been done by mixing the cells or/and growth factors with alginate solution to form gel microspheres in an isotonic $CaCl_2$ solution (**Figure 2**). The findings have shown that the cells are distributed homogeneously inside the gel microspheres. Those cell-containing alginate beads can be used as chondrogenesis-promoting scaffolds for cartilage regeneration [91, 92].

Chitosan is another natural polysaccharide extracted from crustacean shells, particularly from shrimps and crabs. Chitosan contains glucosamine and hyaluronic acid (HA), which are basic components of the native cartilage. Therefore, chitosan is widely used for cartilage tissue engineering. The recent studies have shown that chitosan-hyaluronic acid hydrogel promoted wounded cartilage healing in a rabbit model [93, 94].

Collagen is a main component of the extracellular matrix (ECM) of chondrocytes. Collagen gel has been widely used as substrates for articular cartilage substitutes [95, 96]. Injectable type II collagen gel has been used to treat full-thickness articular cartilage defects [97]. Clinical study has demonstrated that collagen gel can be used to replace cartilage and subchondral bone [98].

Fibrin hydrogels used for articular cartilage repair has been well documented by a review paper [99]. It has been reported that chondrocytes survived in the fibrin gel and enhanced their synthetic activity as evidenced by the increase of the production of GAG and collagen type II [100]. Human fibrin hydrogels have been approved by the Food and Drug Administration (FDA) for cartilage tissue engineering [101].

Hyaluronan is a main component of native cartilage. Similarly to the other native biomaterial scaffold, hyaluronan is the most widely used scaffold for cartilage tissue engineering. The studies have shown that hyaluronan upregulated collagen II expression and downregulated collagen I expression in human MSCs when they were cultured in hyaluronan gel [102].

Although bioactive natural scaffolds have very good biocompatibilities, their mechanical properties still need to be improved. In addition to natural bioactive scaffolds, synthetic materials provide good mechanical properties suitable for cartilage tissue engineering. These synthetic polymers are either used alone or combined with natural biomaterials for cartilage research.

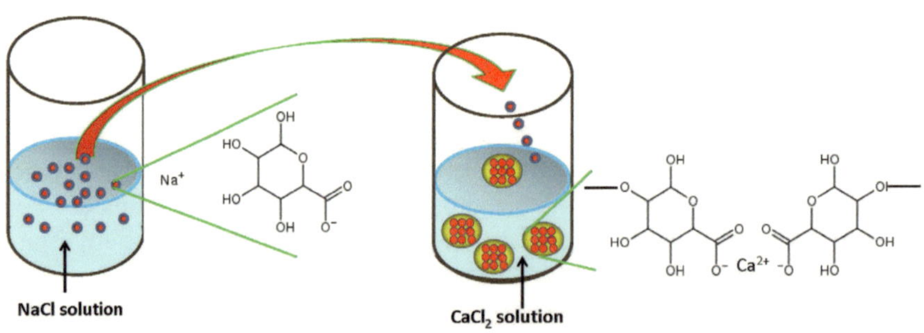

Figure 2.
The intermolecular network of alginate molecules is formed in calcium chloride solution. Alginate can be dissolved with sodium chloride (left image), but cross-linked each other in calcium ions-containing solution to form hydrogel (right image).

The most famous synthetic polymers for cartilage regeneration are polylactic acid (PLA), polyglycolic acid (PGA), and their copolymer polylactic-co-glycolic acid (PLGA). These polymers have a beneficial range of mechanical characteristics and high biocompatibility. Owing to the fact that PLA-PGA polymers have been successfully used in the clinics including sutures, screws, and pins [103–105], they are also used for articular cartilage defect repair in rabbits [106] and meniscal lesion repair in dogs [107]. Currently, two PLA-based scaffolds have been clinically used for cartilage repair: one is BioSeed ®-C and the other one is TRUFIT CB™. The PLA-based polymer scaffolds have shown significant improvement in patient outcomes for the treatment of post-traumatic OA and focal degenerative cartilage defects [108, 109].

Polyethylene glycol (PEG), a nontoxic synthetic polymer, is widely used with other natural materials to enhance their mechanical strength for cartilage tissue engineering. The studies have indicated that PEG-based hydrogel can promote chondrogenic differentiation of MSCs *in vitro* and *in vivo* [110, 111]. Injectable hydrogels used for cartilage tissue engineering have been well summarized by several review papers [112]. PEG-HA scaffold-treated patients achieved significantly higher levels of tissue fill in cartilage defects [113].

5. New surgical techniques for cartilage regeneration

Surgical techniques are more important for cartilage repair. In any cartilage repair techniques, the preparation of the defect bed to receive the implant is essential [114]. In order to promote cartilage regeneration, several new surgical techniques have been developed.

5.1 Arthroscopic surgery

Arthroscopic surgery is a common orthopedic procedure that is used in the diagnosis and treatment of problems inside a joint. Generally, the cartilage defect is measured with an arthroscopic graded probe, and the size and the shape of the defect are templated using sterile paper or aluminum that is subsequently used to prepare the graft if it is not an injectable gel form [114]. Besides the defect preparation and measurement, most operations can be done under an arthroscopy. Currently, arthroscopic surgery has been widely used for various damaged cartilage treatments such as degenerative meniscal tear [115] and osteoarthritis of the knee [116].

5.2 Open surgery

Open surgery is used for some arthroscopically inaccessible cartilage defects such as patella, trochlea, posterior femoral condyle, and some scaffolds that cannot be implanted arthroscopically [114]. This technique has been widely used in cartilage tissue engineering for animal surgery and clinical practice.

5.3 Microfracture surgery

Microfracture surgery is a surgical technique used to repair damaged cartilage by making multiple small holes in the surface of the joint to stimulate a healing response. This technique was developed in the early 1980s by Steadman and his colleagues. The technical details of microfracture have been well summarized [117]. Several animal studies have been completed to assess the microfracture technique [118, 119]. The functional outcomes of patients treated with microfracture for traumatic chondral defects have shown significant improvement [120]. Currently, microfracture surgical technique is considered to be an effective arthroscopic treatment for full-thickness cartilage defect [121]. However, some studies have shown that the younger patients have better clinical outcomes and quality cartilage repair than older patients [122].

5.4 Mosaicplasty surgery

Mosaicplasty surgery is another common cartilage restoration technique in standard clinical practice. This technique was introduced into clinical application in 1992 [123]. Mosaicplasty surgical technique is based on the mosaic-like transplantation of several small, cylindrical plugs of bone and cartilage to provide an even resurfaced area. The long-term clinical follow-up results have shown that the mosaicplasty-treated patients can regain their pre-injury activity level [124].

The studies have demonstrated that the treatment of mosaicplasty in a single cartilage defect size one to five square centimeters of the femoral condyle resulted in clinically relevant better outcome than microfracture [125, 126].

6. Conclusions and future research on cartilage tissue engineering

Cartilage tissue engineering is to use a biomaterial scaffold, bioactive molecules, and cells to produce new cartilage under special conditions. The rapid progress in material science, life science, and engineering has resulted in advancements in the treatment options for various illnesses and diseases, especially for cartilage defects. However, the field of cartilage tissue engineering is still in developing stage. The number of potential variables in cartilage tissue engineering strategy is vast, and the key challenges remain to be addressed. As cartilage tissue engineering incorporates the fields of cell biology, nuclear transfer, and material science, personnel who have mastered the techniques of cell harvest, culture, expansion, transplantation, and polymer design is essential for the successful application of these technologies to build new cartilage and extend human life. The future research on cartilage tissue engineering should thus be aimed at investigating and evaluating tissue engineering approaches, as well as surgical techniques for cartilage repair in disease-compromised animal models to gain a better understanding of clinically feasible design. It is necessary to develop a model system for the study of normal and pathological cartilage tissues.

Conflict of interest

The authors declared no potential conflicts of interest with respect to the research authorship and/or publication of this article.

Funding

This work was financially supported by the Department of Orthopedic Surgery, University of Pittsburgh School of Medicine.

Author details

He Huang[2], Hongyao Xu[2] and Jianying Zhang[1*]

1 Department of Orthopaedic Surgery, University of Pittsburgh School of Medicine, Pittsburgh, PA, USA

2 Department of Sports Medicine and Joint Surgery, Nanjing First Hospital, Nanjing Medical University, Nanjing, Jiangsu, China

*Address all correspondence to: jianying@pitt.edu

References

[1] Sophia Fox AJ, Bedi A, Rodeo SA. The basic science of articular cartilage: Structure, composition, and function. Sports Health. 2009;**1**:461-468

[2] Hootman JM, Helmick CG, Barbour KE, Theis KA, Boring MA. Updated projected prevalence of self-reported doctor-diagnosed arthritis and arthritis-attributable activity limitation among US adults, 2015-2040. Arthritis & Rhematology. 2016;**68**:1582-1587

[3] Helmick CG, Felson DT, Lawrence RC, Gabriel S, Hirsch R, Kwoh CK, et al. Estimates of the prevalence of arthritis and other rheumatic conditions in the United States. Part I. Arthritis and Rheumatism. 2008;**58**:15-25

[4] Paesold G, Nerlich AG, Boos N. Biological treatment strategies for disc degeneration: potentials and shortcomings. European Spine Journal. 2007;**16**:447-468

[5] Walker BF. The prevalence of low back pain: A systematic review of the literature from 1966 to 1998. Journal of Spinal Disorders. 2000;**13**:205-217

[6] Langer R, Vacanti JP. Tissue engineering. Science. 1993;**260**:920-926

[7] Fuchs JR, Nasseri BA, Vacanti JP. Tissue engineering: A 21st century solution to surgical reconstruction. The Annals of Thoracic Surgery. 2001;**72**:577-591

[8] Kessler MW, Grande DA. Tissue engineering and cartilage. Organogenesis. 2008;**4**:28-32

[9] Caplan AI. Review: Mesenchymal stem cells: Cell-based reconstructive therapy in orthopedics. Tissue Engineering. 2005;**11**:1198-1211

[10] Brittberg M, Lindahl A, Nilsson A, Ohlsson C, Isaksson O, Peterson L. Treatment of deep cartilage defects in the knee with autologous chondrocyte transplantation. The New England Journal of Medicine. 1994;**331**:889-895

[11] Horas U, Pelinkovic D, Herr G, Aigner T, Schnettler R. Autologous chondrocyte implantation and osteochondral cylinder transplantation in cartilage repair of the knee joint. A prospective, comparative trial. The Journal of bone and joint surgery American. 2003;**85-A**:185-192

[12] Brittberg M. Cell carriers as the next generation of cell therapy for cartilage repair: A review of the matrix-induced autologous chondrocyte implantation procedure. American Journal of Sports Medicine. 2010;**38**:1259-1271

[13] Peterson L, Vasiliadis HS, Brittberg M, Lindahl A. Autologous chondrocyte implantation: A long-term follow-up. American Journal of Sports Medicine. 2010;**38**:1117-1124

[14] Mardones R, Jofre CM, Minguell JJ. Cell therapy and tissue engineering approaches for cartilage repair and/or regeneration. International Journal of Stem Cells. 2015;**8**:48-53

[15] Mobasheri A, Kalamegam G, Musumeci G, Batt ME. Chondrocyte and mesenchymal stem cell-based therapies for cartilage repair in osteoarthritis and related orthopaedic conditions. Maturitas. 2014;**78**:188-198

[16] Kristjansson B, Honsawek S. Mesenchymal stem cells for cartilage regeneration in osteoarthritis. World Journal of Orthopedics. 2017;**8**:674-680

[17] Friedenstein AJ, Piatetzky S, II, Petrakova KV. Osteogenesis in transplants of bone marrow cells. Journal of Embryology and Experimental Morphology 1966;**16**:381-390

[18] Pittenger MF, Mackay AM, Beck SC, Jaiswal RK, Douglas R, Mosca JD, et al. Multilineage potential of adult human mesenchymal stem cells. Science. 1999;**284**:143-147

[19] Crisan M, Yap S, Casteilla L, Chen CW, Corselli M, Park TS, et al. A perivascular origin for mesenchymal stem cells in multiple human organs. Cell Stem Cell. 2008;**3**:301-313

[20] Chin S, Furukawa K, Ono A, Asari T, Harada Y, Wada K, et al. Immunohistochemical localization of mesenchymal stem cells in ossified human spinal ligaments. Biochemical and Biophysical Research Communications. 2013;**436**:698-704

[21] Schieker M, Gulkan H, Austrup B, Neth P, Mutschler W. Telomerase activity and telomere length of human mesenchymal stem cells. Changes during osteogenic differentiation. Der Orthopäde. 2004;**33**:1373-1377

[22] Li S, Huang KJ, Wu JC, Hu MS, Sanyal M, Hu M, et al. Peripheral blood-derived mesenchymal stem cells: Candidate cells responsible for healing critical-sized calvarial bone defects. Stem Cells Translational Medicine. 2015;**4**:359-368

[23] Sibov TT, Severino P, Marti LC, Pavon LF, Oliveira DM, Tobo PR, et al. Mesenchymal stem cells from umbilical cord blood: Parameters for isolation, characterization and adipogenic differentiation. Cytotechnology. 2012;**64**:511-521

[24] Uccelli A, Moretta L, Pistoia V. Mesenchymal stem cells in health and disease. Nature Reviews. Immunology. 2008;**8**:726-736

[25] Wakitani S, Mitsuoka T, Nakamura N, Toritsuka Y, Nakamura Y, Horibe S. Autologous bone marrow stromal cell transplantation for repair of full-thickness articular cartilage defects in human patellae: Two case reports. Cell Transplantation. 2004;**13**:595-600

[26] Yamasaki S, Mera H, Itokazu M, Hashimoto Y, Wakitani S. Cartilage repair with autologous bone marrow Mesenchymal stem cell transplantation: Review of preclinical and clinical studies. Cartilage. 2014;**5**:196-202

[27] Haleem AM, Singergy AA, Sabry D, Atta HM, Rashed LA, Chu CR, et al. The clinical use of human culture-expanded autologous bone marrow Mesenchymal stem cells transplanted on platelet-rich fibrin glue in the treatment of articular cartilage defects: A pilot study and preliminary results. Cartilage. 2010;**1**:253-261

[28] Wakitani S, Imoto K, Yamamoto T, Saito M, Murata N, Yoneda M. Human autologous culture expanded bone marrow mesenchymal cell transplantation for repair of cartilage defects in osteoarthritic knees. Osteoarthritis and Cartilage. 2002;**10**:199-206

[29] Centeno CJ, Busse D, Kisiday J, Keohan C, Freeman M, Karli D. Regeneration of meniscus cartilage in a knee treated with percutaneously implanted autologous mesenchymal stem cells. Medical Hypotheses. 2008;**71**:900-908

[30] Bielli A, Scioli MG, Gentile P, Cervelli V, Orlandi A. Adipose-derived stem cells in cartilage regeneration: Current perspectives. Regenerative Medicine. 2016;**11**:693-703

[31] Im GI. Regeneration of articular cartilage using adipose stem cells. Journal of Biomedical Materials Research Part A. 2016;**104**:1830-1844

[32] Katz AJ, Llull R, Hedrick MH, Futrell JW. Emerging approaches to the tissue engineering of fat. Clinics in Plastic Surgery. 1999;**26**:587-603

[33] Rodbell M. Metabolism of isolated fat cells. II. The similar effects of phospholipase C (*Clostridium perfringens* alpha toxin) and of insulin on glucose and amino acid metabolism. The Journal of Biological Chemistry. 1966;**241**:130-139

[34] Zhu M, Heydarkhan-Hagvall S, Hedrick M, Benhaim P, Zuk P. Manual isolation of adipose-derived stem cells from human lipoaspirates. Journal of Visualized Experiments. 2013;**79**:e50585

[35] Bunnell BA, Flaat M, Gagliardi C, Patel B, Ripoll C. Adipose-derived stem cells: Isolation, expansion and differentiation. Methods. 2008;**45**:115-120

[36] Peng L, Jia Z, Yin X, Zhang X, Liu Y, Chen P, et al. Comparative analysis of mesenchymal stem cells from bone marrow, cartilage, and adipose tissue. Stem Cells and Development. 2008;**17**:761-773

[37] Cui L, Wu Y, Cen L, Zhou H, Yin S, Liu G, et al. Repair of articular cartilage defect in non-weight bearing areas using adipose derived stem cells loaded polyglycolic acid mesh. Biomaterials. 2009;**30**:2683-2693

[38] Hatakeyama A, Uchida S, Utsunomiya H, Tsukamoto M, Nakashima H, Nakamura E, et al. Isolation and characterization of synovial Mesenchymal stem cell derived from hip joints: A comparative analysis with a matched control knee group. Stem Cells International. 2017;**2017**:9312329

[39] De Bari C, Dell'Accio F, Tylzanowski P, Luyten FP. Multipotent mesenchymal stem cells from adult human synovial membrane. Arthritis and Rheumatism. 2001;**44**:1928-1942

[40] Yao Y, Li ZY, Zhang H, Zheng YH, Mai LX, Liu WJ, et al. Synovial fluid-derived synovial fragments represent an improved source of synovial mesenchymal stem cells in the temporomandibular joint. International Journal of Molecular Medicine. 2018;**41**:173-183

[41] Benya PD, Shaffer JD. Dedifferentiated chondrocytes reexpress the differentiated collagen phenotype when cultured in agarose gels. Cell. 1982;**30**:215-224

[42] Zayed M, Newby S, Misk N, Donnell R, Dhar M. Xenogenic implantation of equine synovial fluid-derived Mesenchymal stem cells leads to articular cartilage regeneration. Stem Cells International. 2018;**2018**:1073705

[43] Niemeyer P, Schonberger TS, Hahn J, Kasten P, Fellenberg J, Suedkamp N, et al. Xenogenic transplantation of human mesenchymal stem cells in a critical size defect of the sheep tibia for bone regeneration. Tissue Engineering Parts A. 2010;**16**:33-43

[44] Henriksson HB, Svanvik T, Jonsson M, Hagman M, Horn M, Lindahl A, et al. Transplantation of human mesenchymal stems cells into intervertebral discs in a xenogeneic porcine model. Spine. 2009;**34**:141-148

[45] Fortier LA, Barker JU, Strauss EJ, McCarrel TM, Cole BJ. The role of growth factors in cartilage repair. Clinical Orthopaedics and Related Research. 2011;**469**:2706-2715

[46] Chubinskaya S, Segalite D, Pikovsky D, Hakimiyan AA, Rueger DC. Effects induced by BMPS in cultures of human articular chondrocytes: Comparative studies. Growth Factors. 2008;**26**:275-283

[47] Davidson TS, DiPaolo RJ, Andersson J, Shevach EM. Cutting edge: IL-2 is essential for TGF-beta-mediated induction of Foxp3+ T regulatory cells. Journal of Immunology. 2007;**178**:4022-4026

[48] Lee CH, Cook JL, Mendelson A, Moioli EK, Yao H, Mao JJ. Regeneration of the articular surface of the rabbit synovial joint by cell homing: A proof of concept study. Lancet. 2010;**376**:440-448

[49] Blaney Davidson EN, Vitters EL, van Lent PL, van de Loo FA, van den Berg WB, van der Kraan PM. Elevated extracellular matrix production and degradation upon bone morphogenetic protein-2 (BMP-2) stimulation point toward a role for BMP-2 in cartilage repair and remodeling. Arthritis Research & Therapy. 2007;**9**:R102

[50] Chubinskaya S, Hurtig M, Rueger DC. OP-1/BMP-7 in cartilage repair. International Orthopaedics. 2007;**31**:773-781

[51] Miyamoto C, Matsumoto T, Sakimura K, Shindo H. Osteogenic protein-1 with transforming growth factor-beta1: Potent inducer of chondrogenesis of synovial mesenchymal stem cells in vitro. Journal of Orthopaedic Science. 2007;**12**:555-561

[52] Shen B, Wei A, Whittaker S, Williams LA, Tao H, Ma DD, et al. The role of BMP-7 in chondrogenic and osteogenic differentiation of human bone marrow multipotent mesenchymal stromal cells in vitro. Journal of Cellular Biochemistry. 2010;**109**:406-416

[53] Ekenstedt KJ, Sonntag WE, Loeser RF, Lindgren BR, Carlson CS. Effects of chronic growth hormone and insulin-like growth factor 1 deficiency on osteoarthritis severity in rat knee joints. Arthritis and Rheumatism. 2006;**54**:3850-3858

[54] Mullen LM, Best SM, Ghose S, Wardale J, Rushton N, Cameron RE. Bioactive IGF-1 release from collagen-GAG scaffold to enhance cartilage repair in vitro. Journal of Materials Science Materials in Medicine. 2015;**26**:5325

[55] Tyler JA. Insulin-like growth factor 1 can decrease degradation and promote synthesis of proteoglycan in cartilage exposed to cytokines. Biochemical Journal. 1989;**260**:543-548

[56] Zhang M, Zhou Q, Liang QQ, Li CG, Holz JD, Tang D, et al. IGF-1 regulation of type II collagen and MMP-13 expression in rat endplate chondrocytes via distinct signaling pathways. Osteoarthritis and Cartilage. 2009;**17**:100-106

[57] Ortved KF, Begum L, Mohammed HO, Nixon AJ. Implantation of rAAV5-IGF-I transduced autologous chondrocytes improves cartilage repair in full-thickness defects in the equine model. Molecular Therapy. 2015;**23**:363-373

[58] Crane JL, Cao X. Function of matrix IGF-1 in coupling bone resorption and formation. Journal of Molecular Medicine. 2014;**92**:107-115

[59] Massicotte F, Fernandes JC, Martel-Pelletier J, Pelletier JP, Lajeunesse D. Modulation of insulin-like growth factor 1 levels in human osteoarthritic subchondral bone osteoblasts. Bone. 2006;**38**:333-341

[60] Cuevas P, Burgos J, Baird A. Basic fibroblast growth factor (FGF) promotes cartilage repair in vivo. Biochemical and Biophysical Research Communications. 1988;**156**:611-618

[61] Im HJ, Muddasani P, Natarajan V, Schmid TM, Block JA, Davis F, et al. Basic fibroblast growth factor stimulates matrix metalloproteinase-13 via the molecular cross-talk between the mitogen-activated protein kinases and protein kinase Cdelta pathways in human adult articular chondrocytes. The Journal of Biological Chemistry. 2007;**282**:11110-11121

[62] Yamamoto T, Wakitani S, Imoto K, Hattori T, Nakaya H, Saito M, et al.

Fibroblast growth factor-2 promotes the repair of partial thickness defects of articular cartilage in immature rabbits but not in mature rabbits. Osteoarthritis and Cartilage. 2004;**12**:636-641

[63] Moore EE, Bendele AM, Thompson DL, Littau A, Waggie KS, Reardon B, et al. Fibroblast growth factor-18 stimulates chondrogenesis and cartilage repair in a rat model of injury-induced osteoarthritis. Osteoarthritis and Cartilage. 2005;**13**:623-631

[64] Mori Y, Saito T, Chang SH, Kobayashi H, Ladel CH, Guehring H, et al. Identification of fibroblast growth factor-18 as a molecule to protect adult articular cartilage by gene expression profiling. The Journal of Biological Chemistry. 2014;**289**:10192-10200

[65] Huang L, Yi L, Zhang C, He Y, Zhou L, Liu Y, et al. Synergistic effects of FGF-18 and TGF-beta3 on the Chondrogenesis of human adipose-derived Mesenchymal stem cells in the pellet culture. Stem Cells International. 2018;**2018**:7139485

[66] Barr L, Getgood A, Guehring H, Rushton N, Henson FM. The effect of recombinant human fibroblast growth factor-18 on articular cartilage following single impact load. Journal of Orthopaedic Research. 2014;**32**:923-927

[67] Okuda K, Kawase T, Momose M, Murata M, Saito Y, Suzuki H, et al. Platelet-rich plasma contains high levels of platelet-derived growth factor and transforming growth factor-beta and modulates the proliferation of periodontally related cells in vitro. Journal of Periodontology. 2003;**74**:849-857

[68] Lubkowska A, Dolegowska B, Banfi G. Growth factor content in PRP and their applicability in medicine. Journal of Biological Regulators and Homeostatic Agents. 2012;**26**:3S-22S

[69] Hall MP, Band PA, Meislin RJ, Jazrawi LM, Cardone DA. Platelet-rich plasma: Current concepts and application in sports medicine. Journal of the American Academy of Orthopaedic Surgeons. 2009;**17**:602-608

[70] Zhang J, Yuan T, Zheng N, Zhou Y, Hogan MV, Wang JH. The combined use of kartogenin and platelet-rich plasma promotes fibrocartilage formation in the wounded rat Achilles tendon entheses. Bone & Joint Research. 2017;**6**:231-244

[71] Zhang J, Wang JH. Kartogenin induces cartilage-like tissue formation in tendon-bone junction. Bone Research. 2014;**2**:14008

[72] Johnson K, Zhu S, Tremblay MS, Payette JN, Wang J, Bouchez LC, et al. A stem cell-based approach to cartilage repair. Science. 2012;**336**:717-721

[73] Huang H, Xu H, Zhao J. A novel approach for meniscal regeneration using Kartogenin-treated autologous tendon graft. American Journal of Sports Medicine. 2017;**45**:3289-3297

[74] Zhou Y, Zhang J, Yang J, Narava M, Zhao G, Yuan T, et al. Kartogenin with PRP promotes the formation of fibrocartilage zone in the tendon-bone interface. Journal of Tissue Engineering and Regenerative Medicine. 2017;**11**:3445-3456

[75] Huang Y, Jiang T, Chen J, Yin GY, Fan J. Effects of kartogenin on the attenuated nucleus pulposus cell degeneration of intervertebral discs induced by interleukin-1beta and tumor necrosis factor-alpha. International Journal of Molecular Medicine. 2018;**41**:749-756

[76] Kwon JY, Lee SH, Na HS, Jung K, Choi J, Cho KH, et al. Kartogenin inhibits pain behavior, chondrocyte inflammation, and attenuates osteoarthritis progression in mice

through induction of IL-10. Scientific Reports. 2018;**8**:13832

[77] Wang D, Tan H, Lebaschi AH, Nakagawa Y, Wada S, Donnelly PE, et al. Kartogenin enhances collagen organization and mechanical strength of the repaired Enthesis in a murine model of rotator cuff repair. Arthroscopy. 2018;**34**:2579-2587

[78] Westin CB, Trinca RB, Zuliani C, Coimbra IB, Moraes AM. Differentiation of dental pulp stem cells into chondrocytes upon culture on porous chitosan-xanthan scaffolds in the presence of kartogenin. Materials Science and Engineering: C. 2017;**80**:594-602

[79] Mobasheri A. The future of osteoarthritis therapeutics: Emerging biological therapy. Current Rheumatology Reports. 2013;**15**:385

[80] Smith DA, Galin I. Statin therapy for native and peri-interventional coronary heart disease. Current Molecular Medicine. 2006;**6**:589-602

[81] Lazzerini PE, Capecchi PL, Selvi E, Lorenzini S, Bisogno S, Baldari CT, et al. Statins and the joint: Multiple targets for a global protection? Seminars in Arthritis and Rheumatism. 2011;**40**:430-446

[82] Leung BP, Sattar N, Crilly A, Prach M, McCarey DW, Payne H, et al. A novel anti-inflammatory role for simvastatin in inflammatory arthritis. Journal of Immunology. 2003;**170**:1524-1530

[83] Dombrecht EJ, Van Offel JF, Bridts CH, Ebo DG, Seynhaeve V, Schuerwegh AJ, et al. Influence of simvastatin on the production of pro-inflammatory cytokines and nitric oxide by activated human chondrocytes. Clinical and Experimental Rheumatology. 2007;**25**:534-539

[84] Akasaki Y, Matsuda S, Nakayama K, Fukagawa S, Miura H, Iwamoto Y. Mevastatin reduces cartilage degradation in rabbit experimental osteoarthritis through inhibition of synovial inflammation. Osteoarthritis and Cartilage. 2009;**17**:235-243

[85] Oka S, Matsumoto T, Kubo S, Matsushita T, Sasaki H, Nishizawa Y, et al. Local administration of low-dose simvastatin-conjugated gelatin hydrogel for tendon-bone healing in anterior cruciate ligament reconstruction. Tissue Engineering Parts A. 2013;**19**:1233-1243

[86] Awad HA, Wickham MQ, Leddy HA, Gimble JM, Guilak F. Chondrogenic differentiation of adipose-derived adult stem cells in agarose, alginate, and gelatin scaffolds. Biomaterials. 2004;**25**:3211-3222

[87] Weisser J, Rahfoth B, Timmermann A, Aigner T, Brauer R, von der Mark K. Role of growth factors in rabbit articular cartilage repair by chondrocytes in agarose. Osteoarthritis and Cartilage. 2001;**9**(Suppl A):S48-S54

[88] Yin Z, Yang X, Jiang Y, Xing L, Xu Y, Lu Y, et al. Platelet-rich plasma combined with agarose as a bioactive scaffold to enhance cartilage repair: An in vitro study. Journal of Biomaterials Applications. 2014;**28**:1039-1050

[89] Sun J, Tan H. Alginate-based biomaterials for regenerative medicine applications. Materials. 2013;**6**:1285-1309

[90] Mata M, Milian L, Oliver M, Zurriaga J, Sancho-Tello M, de Llano JJM, et al. In vivo articular cartilage regeneration using human dental pulp stem cells cultured in an alginate scaffold: A preliminary study. Stem Cells International. 2017;**2017**:8309256

[91] Focaroli S, Teti G, Salvatore V, Orienti I, Falconi M. Calcium/cobalt alginate beads as functional scaffolds for cartilage tissue engineering. Stem Cells International. 2016;**2016**:2030478

[92] Pleumeekers MM, Nimeskern L, Koevoet WL, Kops N, Poublon RM, Stok KS, et al. The in vitro and in vivo capacity of culture-expanded human cells from several sources encapsulated in alginate to form cartilage. European Cells & Materials. 2014;**27**:264-280; discussion 78-80

[93] Mohan N, Mohanan PV, Sabareeswaran A, Nair P. Chitosan-hyaluronic acid hydrogel for cartilage repair. International Journal of Biological Macromolecules. 2017;**104**:1936-1945

[94] Zhao M, Chen Z, Liu K, Wan YQ, Li XD, Luo XW, et al. Repair of articular cartilage defects in rabbits through tissue-engineered cartilage constructed with chitosan hydrogel and chondrocytes. Journal of Zhejiang University Science B. 2015;**16**:914-923

[95] Mueller-Rath R, Gavenis K, Andereya S, Mumme T, Albrand M, Stoffel M, et al. Condensed cellular seeded collagen gel as an improved biomaterial for tissue engineering of articular cartilage. Bio-Medical Materials and Engineering. 2010;**20**:317-328

[96] Kawamura S, Wakitani S, Kimura T, Maeda A, Caplan AI, Shino K, et al. Articular cartilage repair. Rabbit experiments with a collagen gel-biomatrix and chondrocytes cultured in it. Acta Orthopaedica Scandinavica. 1998;**69**:56-62

[97] Funayama A, Niki Y, Matsumoto H, Maeno S, Yatabe T, Morioka H, et al. Repair of full-thickness articular cartilage defects using injectable type II collagen gel embedded with cultured chondrocytes in a rabbit model. Journal of Orthopaedic Science. 2008;**13**:225-232

[98] Kon E, Delcogliano M, Filardo G, Busacca M, Di Martino A, Marcacci M. Novel nano-composite multilayered biomaterial for osteochondral regeneration: A pilot clinical trial. American Journal of Sports Medicine. 2011;**39**:1180-1190

[99] Peretti GM, Xu JW, Bonassar LJ, Kirchhoff CH, Yaremchuk MJ, Randolph MA. Review of injectable cartilage engineering using fibrin gel in mice and swine models. Tissue Engineering. 2006;**12**:1151-1168

[100] Scotti C, Mangiavini L, Boschetti F, Vitari F, Domeneghini C, Fraschini G, et al. Effect of in vitro culture on a chondrocyte-fibrin glue hydrogel for cartilage repair. Knee Surgery, Sports Traumatology, Arthroscopy. 2010;**18**:1400-1406

[101] Fussenegger M, Meinhart J, Hobling W, Kullich W, Funk S, Bernatzky G. Stabilized autologous fibrin-chondrocyte constructs for cartilage repair in vivo. Annals of Plastic Surgery. 2003;**51**:493-498

[102] Lisignoli G, Cristino S, Piacentini A, Toneguzzi S, Grassi F, Cavallo C, et al. Cellular and molecular events during chondrogenesis of human mesenchymal stromal cells grown in a three-dimensional hyaluronan based scaffold. Biomaterials. 2005;**26**:5677-5686

[103] Bostman O, Makela EA, Sodergard J, Hirvensalo E, Tormala P, Rokkanen P. Absorbable polyglycolide pins in internal fixation of fractures in children. Journal of Pediatric Orthopedics. 1993;**13**:242-245

[104] Hirvensalo E, Bostman O, Partio E, Tormala P, Rokkanen P. Fracture of the humeral capitellum fixed with absorbable polyglycolide pins. 1-year follow-up of 8 adults. Acta Orthopaedica Scandinavica. 1993;**64**:85-86

[105] Paivarinta U, Bostman O, Majola A, Toivonen T, Tormala P, Rokkanen P. Intraosseous cellular response to

biodegradable fracture fixation screws made of polyglycolide or polylactide. Archives of Orthopaedic and Trauma Surgery. 1993;**112**:71-74

[106] von Schroeder HP, Kwan M, Amiel D, Coutts RD. The use of polylactic acid matrix and periosteal grafts for the reconstruction of rabbit knee articular defects. Journal of Biomedical Materials Research. 1991;**25**:329-339

[107] Klompmaker J, Jansen HW, Veth RP, de Groot JH, Nijenhuis AJ, Pennings AJ. Porous polymer implant for repair of meniscal lesions: A preliminary study in dogs. Biomaterials. 1991;**12**:810-816

[108] Kreuz PC, Muller S, Ossendorf C, Kaps C, Erggelet C. Treatment of focal degenerative cartilage defects with polymer-based autologous chondrocyte grafts: Four-year clinical results. Arthritis Research & Therapy. 2009;**11**:R33

[109] Kreuz PC, Muller S, Freymann U, Erggelet C, Niemeyer P, Kaps C, et al. Repair of focal cartilage defects with scaffold-assisted autologous chondrocyte grafts: Clinical and biomechanical results 48 months after transplantation. American Journal of Sports Medicine. 2011;**39**:1697-1705

[110] Skaalure SC, Chu S, Bryant SJ. An enzyme-sensitive PEG hydrogel based on aggrecan catabolism for cartilage tissue engineering. Advanced Healthcare Materials. 2015;**4**:420-431

[111] Wang J, Zhang F, Tsang WP, Wan C, Wu C. Fabrication of injectable high strength hydrogel based on 4-arm star PEG for cartilage tissue engineering. Biomaterials. 2017;**120**:11-21

[112] Liu M, Zeng X, Ma C, Yi H, Ali Z, Mou X, et al. Injectable hydrogels for cartilage and bone tissue engineering. Bone Research. 2017;**5**:17014

[113] Sharma B, Fermanian S, Gibson M, Unterman S, Herzka DA, Cascio B, et al. Human cartilage repair with a photoreactive adhesive-hydrogel composite. Science Translational Medicine. 2013;**5**:167ra6

[114] Haleem AM, Chu CR. Advances in tissue engineering techniques for articular cartilage repair. Operative Techniques in Orthopaedics. 2010;**20**:76-89

[115] Sihvonen R, Paavola M, Malmivaara A, Itala A, Joukainen A, Nurmi H, et al. Arthroscopic partial meniscectomy versus sham surgery for a degenerative meniscal tear. The New England Journal of Medicine. 2013;**369**:2515-2524

[116] Kirkley A, Birmingham TB, Litchfield RB, Giffin JR, Willits KR, Wong CJ, et al. A randomized trial of arthroscopic surgery for osteoarthritis of the knee. The New England Journal of Medicine. 2008;**359**:1097-1107

[117] Steadman JR, Rodkey WG, Briggs KK. Microfracture: Its history and experience of the developing surgeon. Cartilage. 2010;**1**:78-86

[118] Frisbie DD, Morisset S, Ho CP, Rodkey WG, Steadman JR, McIlwraith CW. Effects of calcified cartilage on healing of chondral defects treated with microfracture in horses. American Journal of Sports Medicine. 2006;**34**:1824-1831

[119] Frisbie DD, Oxford JT, Southwood L, Trotter GW, Rodkey WG, Steadman JR, et al. Early events in cartilage repair after subchondral bone microfracture. Clinical Orthopaedics and Related Research. 2003;**407**:215-227

[120] Steadman JR, Briggs KK, Rodrigo JJ, Kocher MS, Gill TJ, Rodkey WG. Outcomes of microfracture for traumatic chondral defects of the knee: Average 11-year follow-up. Arthroscopy. 2003;**19**:477-484

[121] Miller BS, Steadman JR, Briggs KK, Rodrigo JJ, Rodkey WG. Patient satisfaction and outcome after microfracture of the degenerative knee. The Journal of Knee Surgery. 2004;**17**:13-17

[122] Kreuz PC, Erggelet C, Steinwachs MR, Krause SJ, Lahm A, Niemeyer P, et al. Is microfracture of chondral defects in the knee associated with different results in patients aged 40 years or younger? Arthroscopy. 2006;**22**:1180-1186

[123] Hangody L, Kish G, Karpati Z, Szerb I, Udvarhelyi I. Arthroscopic autogenous osteochondral mosaicplasty for the treatment of femoral condylar articular defects. A preliminary report. Knee Surgery, Sports Traumatology, Arthroscopy. 1997;**5**:262-267

[124] Cognault J, Seurat O, Chaussard C, Ionescu S, Saragaglia D. Return to sports after autogenous osteochondral mosaicplasty of the femoral condyles: 25 cases at a mean follow-up of 9 years. Orthopaedics & Traumatology: Surgery & Research. 2015;**101**:313-317

[125] Solheim E, Hegna J, Inderhaug E. Long-term clinical follow-up of microfracture versus mosaicplasty in articular cartilage defects of medial femoral condyle. The Knee. 2017;**24**:1402-1407

[126] Solheim E, Hegna J, Strand T, Harlem T, Inderhaug E. Randomized study of long-term (15-17 years) outcome after microfracture versus Mosaicplasty in knee articular cartilage defects. American Journal of Sports Medicine. 2018;**46**:826-831

Chapter 7

Cartilage Tissue Engineering Using Self-Assembling Peptides Composite Scaffolds

Nausika Betriu and Carlos E. Semino

Abstract

Adult articular cartilage presents poor intrinsic capacity for regeneration, and after injury, cellular or biomaterial-based therapeutic platforms are required to assist repair promotion. Cartilage tissue engineering (CTE) aims to produce cartilage-like tissues that recreate the complex mechanical, biophysical and biological properties found *in vivo*. In terms of biomaterials used for CTE, three-dimensional (3D) self-assembling peptide scaffolds (SAPS) are very attractive for their unique properties, such as biocompatibility, optional possibility of rationally design cell-signaling capacity, biodegradability and modulation of its biomechanical properties. The most attractive cell types currently used for CTE are autologous chondrocytes and adult stem cells. The use of chondrocytes in cell-based therapies for cartilage lesions is limited by quantity and requires an *in vitro* 2D expansion, which leads to cell dedifferentiation. In the present chapter, we report the development of heparin-, chondroitin sulfate-, decorin-, and poly(ε-caprolactone)-based self-assembling peptide composite scaffolds to promote re-differentiation of expanded human articular chondrocytes and induction of adipose-derived stem cells to chondrogenic commitment.

Keywords: 3D cell culture, cartilage, self-assembling peptide scaffold, biomimetic materials, tissue engineering

1. Introduction

Articular cartilage is an avascular connective tissue, composed of chondrocytes as practically unique cell type. Articular chondrocytes synthesize, maintain and remodel the highly specialized extracellular matrix (ECM) [1], which in turn allows to withstand the mechanical requirements of the joints [2]. It is currently believed that due to its avascular nature, cartilage tissue lacks an intrinsic capacity for regeneration in response to disease or injury, leading to long-term pain, degeneration and loss of function [3]. Cartilage tissue engineering (CTE) aims to produce cartilage-like tissue substitutes by combining the appropriate cells, scaffolds and bioactive molecules to assist repair cartilage lesions [4, 5].

Cell types currently used for CTE include autologous articular chondrocytes (ACh), which already possess the desired phenotype, and mesenchymal stem cells (MSC), from bone marrow (BMSC) or adipose tissue-derived (ADSC), which can be induced to undergo chondrogenic differentiation [6, 7]. Autologous chondrocytes would be the ideal cell source for cartilage repair due to their intrinsic properties regarding cell function and immune compatibility. However, cell accessibility from a patient biopsy is limited, and once isolated, chondrocytes needs to be extensively expanded in 2D monolayer [1]. During expansion process, chondrocytes rapidly undergo extensive loss of the original tissue-specific phenotype, downregulating the expression of chondrogenic markers, such as collagens and glycosaminoglycans while acquiring a fibroblast-like phenotype [8, 9].

Three-dimensional (3D) culture platforms are currently used to restore or maintain chondrogenic phenotype, since it recreates more closely the complex cellular microenvironment found *in vivo* [10, 11]. In terms of biomaterials used for CTE diverse possibilities in composition, structure, biodegradability and biomechanical properties exist. In general, biomaterials user for tissue engineering applications can be classified into natural or synthetic scaffolds. Natural scaffolds are commonly hydrogels made of natural materials such as Matrigel™, collagen type I, laminin and gelatin, which provide chemical cues, principally ECM binding motifs. However, due to its natural origin, they frequently contain undefined amounts of different constituents such as growth factors and cytokines which would be the main responsible of presenting variability from batch to batch [10]. Thus, due to its complex composition possible modifications to improve them are limited. On the other hand, synthetic scaffolds have minimal variation from batch to batch production, providing a reproducible cellular microenvironment. Moreover, they present lower biodegradability *in vitro*, fact that permits to maintain structural and mechanical properties for longer periods of time. Alike natural scaffolds, structural properties, such as matrix stiffness, can be modulated by increasing concentration. In the last decades, polymeric scaffolds, such as poly(lactic-co-glycolic acid) (PLGA) [12] and polylactic acid (PLA) [13, 14] as well as synthetic peptide nanofibers [15] have been developed to culture cells in 3D. Clinically used scaffolds are collagen type I/III and hyaluronic acid-based biomaterials, and others under consideration are for instance injectable fibrin gels, collagen type I or II and sponges, polylactic acid (PLA) and polyglycolic acid (PGA). As today, however, the best CTE product does not maintain their tissue properties after implantation, and the minimal medical standards are not yet achieved.

Synthetic hydrogels are good candidates for CTE since they possess unique properties, such as more than 95% of water content (which mimics the native cartilage ECM), biocompatibility and capacity of rationally design chemical signaling and biochemical properties. One of the best examples is the self-assembling peptide scaffold RAD16-I, commercially available as Puramatrix™. RAD16-I is a short peptide constituted by the sequence AcN-$(RADA)_4$-$CONH_2$, which alternates hydrophilic and hydrophobic amino acids (**Figure 1A**) [16]. The peptide undergoes self-assembly into a nanofiber network with antiparallel β-sheet configuration under physiological conditions (**Figure 1B**) [17]. The nanoscale architecture of the fiber network (around 10 nm diameter and 50–200 nm pore size) allows the cells to experiment a truly 3D environment (**Figure 1C**). Besides, biomolecules in such nanoscale environment diffuse slowly and are likely to establish a local molecular gradient. Non-covalent interactions allow cell growth, migration, contact with other cells, shape changes and a properly exposition of membrane receptors. Moreover, since stiffness can be controlled by changing peptide concentration these hydrogels can be tuned up to embed cells but not to entrap them [18].

Since the peptide scaffold does not contain signaling motifs, the environment can be considered non-instructive, from the point of view of cell receptor recognition/activation. However, the self-assembling peptide scaffold RAD16-I can be functionalized by solid-phase synthesis by extending at the N-termini with signaling motifs, such as ECM ligands for cell receptors, to trigger different cellular

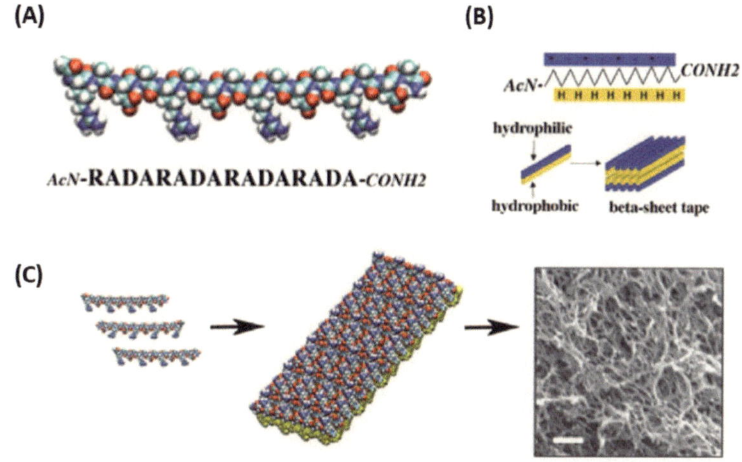

Figure 1.
Peptide RAD16-I self-assembles into a nanofiber network. (A) Molecular model of peptide RAD16-I. Since the scaffold contains no signaling moieties, the environment is not instructive for cells. R = Arg; A = Ala; D = Asp. (B) Molecular model of the nanofiber developed by self-assembling RAD16-I molecules. The nanofiber is formed by a double tape of assembled RAD16-I molecules in antiparallel β-sheet configuration. (C) RAD16-I nanofiber network viewed by SEM. The nanoscale architecture of the fiber network) allows the cells to experiment a truly 3D environment white bar represents 200 nm. Adapted from Semino [17].

responses [16, 19]. Several studies showed the capacity of RAD16-I to support cell maintenance of multiple cell types, including endothelial cells [20], hepatocytes [19, 21], fibroblasts [22], embryonic [23] and somatic stem cells [24, 25].

In the present chapter, we report the development of new bicomponent scaffolds based on the self-assembling peptide RAD16-I, for guiding chondrogenic differentiation of both adipose-derived stem cells (ADSC) and expanded dedifferentiated human articular chondrocytes (hAChs).

On one hand, we took advantage of the versatility of RAD16-I to specifically add molecular cues for guiding chondrogenesis in order to develop more biomimetic scaffolds. Thus, the first approach was based on the addition of heparin (Hep) moieties to the peptide scaffold, forming a stable electrostatic-based composite made of heparin-self-assembling peptide hydrogel. The advantage of this bicomponent scaffold is its natural capacity to retain heparin binding domain (HBD) containing growth factors (GFs), and thus, protecting them from degradation or denaturation [25]. Therefore, the non-instructive RAD16-I scaffold provides the structural 3D environment while the heparin moiety the binding structure to HBD-containing GFs. Our second approach, was based on mimicking the native articular cartilage ECM while providing signaling moieties presented in mature cartilage. Glycosaminoglycans (GAGs) and proteoglycans (PGs) are structural components of the native cartilage ECM and influence the regulation of cell proliferation, migration and differentiation [26]. In particular, chondroitin sulfate (CS, a sulfated GAG usually found as a constituent of PGs) and decorin (a small PG, consisting of a protein core linked to a GAG chain, consisting of chondroitin sulfate or dermatan sulfate) [27, 28] molecules were added to the RAD16-I scaffold by mixing the components, obtaining a chondroitin sulfate- and decorin-based self-assembling peptide composite scaffold.

Finally, we combined the self-assembling peptide RAD16-I with a woven poly (ε-caprolactone) (PCL). 3D weaving can be used to create porous structures arranged in multiple layers of continuous fibers in three orthogonal directions [29]. Such scaffolds were engineered with predetermined properties aiming to reproduce the mechanical features of native articular cartilage. Moreover, PCL is a Food and Drug

Administration (FDA) approved biomaterial, biocompatible and biodegradable, widely used for medical applications [30, 31]. Our strategy was based on combine these two biomaterials to promote the attachment and differentiation of embedded cells, providing at the same time a biomimetic mechanical environment of the native mature cartilage [32].

2. Materials and methods

2.1 2D culture of ADSC and hACh

ADSC (PT-5006, Lonza) were cultured in 175 cm^2 T-flasks (<6th passage) in Adipose-Derived Stem Cell Basal Medium (ADSC-BM) (PT-3273, Lonza) supplemented with Adipose-Derived Stem Cell Growth Medium (ADSC-GM) SingleQuots (PT-4503, Lonza). hACh cells (CC-2550, Lonza) were cultured at 10,000 cells/cm^2 from passages 2–6 in 25, 75, and 175 cm^2 T flasks. The growth medium consisted of Chondrocyte Basal Medium (CBM) (CC-3217, Lonza) plus SingleQuots of Growth Supplements (CC-4409, Lonza) containing R3-IGF-1, bFGF, transferrin, insulin, FBS, and gentamicin/amphotericin-B. Cultures were maintained in the incubator in humidified atmosphere at 37°C and 5% CO_2.

2.2 3D culture of ADSC and hACh in RAD16-I composites scaffolds

ADSC and hACh 3D cultures were maintained under control or chondrogenic conditions. Control medium was prepared with DMEM High Glucose, GlutaMAX (61965, Gibco), ITS + Premix 100x (354352, BDBioscience), 100 U/mL Penicillin/100 µg/mL Streptomycin (P11-010, PAA), 40 µg/mL L-Proline (P5607, Sigma) and 1 mM Sodium Pyruvate (11360, Life Technologies). Cultures for chondrogenic differentiation were induced at day 2 with chondrogenic medium (control medium supplemented with 10 ng/mL TGFβ1 (GF111, Millipore), 25 µg/mL L-ascorbic acid 2-phosphate (A8960; Sigma) and 100 nM Dexamethasone (D8893; Sigma)). Chondrocytes were also cultured in expansion medium (see Section 2.1). 3D cell cultures were maintained in the incubator at 37°C and 5% CO_2, and medium was changed every other day. Cultures were maintained for 4 weeks in the described serum-free media under control or chondrogenic conditions (in the presence of stimulating factors to induce chondrogenic differentiation) [33, 34]. After 4 weeks, 3D constructs were analyzed for morphology, gene and protein expression, glycosaminoglycans production and mechanical properties.

2.3 ADSC 3D culture in RAD/heparin composite scaffold

RAD16-I (PuraMatrix™, 354250, Corning) and composites RAD/Hep were prepared at a final concentration of 0.3% (w/v) RAD16-I. The composites were prepared by combining 95 µL of 0.5% (w/v) RAD16-I and 5 µL of heparin sodium salt solution (H3149, Sigma) in a concentration range between 0.01% and 1% (w/v). The mixture was then diluted with 10% sucrose (S0389, Sigma) to a final concentration of 0.3% (w/v) RAD16-I. To obtain RAD and RAD/Hep 3D cultures, ADSC were harvested by trypsinization and resuspended to 4×10^6 cells/mL in 10% sucrose. The 0.3% (w/v) RAD16-I peptide solution was mixed with the cell suspension (1:1) to obtain a final concentration of 0.15% (w/v) RAD16-I and 2×10^6 cells/mL. Then, 80 µL of the cell-peptide mixture (160,000 cells) was loaded into individual wells of a 48-well culture containing 150 µL of medium, which induced the self-assembly of the peptide. The plate was placed in the incubator for 20 min at 37°C and 5% CO_2, and then 650 μL of fresh medium was added to the 3D cell cultures. ADSC 3D cultures were maintained during 2 days under control medium. Cultures for chondrogenic differentiation were induced at day 2 with chondrogenic medium.

2.3.1 ADSC and hACh 3D culture in RAD/CS and RAD/Dec composite scaffold

RAD16-I (PuraMatrix™, 354250, Corning) and composites RAD/CS and RAD/Decorin were prepared at a final concentration of 0.3% (w/v) RAD16-I. The composites were prepared by combining 95 μL of 0.5% (w/v) RAD16-I and 5 μL CS or Decorin at a concentration of 0.2% (w/v). The mixture was then diluted with 10% sucrose (S0389, Sigma) to a final concentration of 0.3% (w/v) RAD16-I. To obtain RAD16-I, RAD/CS and RAD/Decorin 3D cultures, cells were harvested by trypsinization and resuspended to 4×10^6 cells/mL in 10% sucrose. The 0.3% (w/v) RAD16-I peptide solution was mixed with the cell suspension (1:1) to obtain a final concentration of 0.15% (w/v) RAD16-I and 2×10^6 cells/mL. Then, 80 μL of the cell-peptide mixture (160,000 cells) was loaded into individual wells of a 48-well culture containing 150 μL of control or expansion medium, which induced the self-assembly of the peptide. The plate was placed in the incubator for 20 min at 37°C and 5% CO_2, and then 650 μL of fresh medium was added to the 3D cell cultures.

2.3.2 hACh 3D culture in PCL, PCL/RAD and RAD scaffolds

In the case of PCL scaffold, a cell suspension of 25×10^6 cells/mL was seeded onto the surface of 5 mm × 0.75 mm woven PCL scaffolds (500,000 cells/scaffold). After 2 h, 100 μL of expansion or control medium were slowly added into the well and after 4 h, 700 μL were finally added. For PCL/RAD composites, cells were harvested and resuspended to 50×10^6 cells/mL in 10% (w/v) sucrose. Then, cells were equally mixed with 1% (w/v) RAD16-I and seeded onto the woven PCL scaffold disks (500,000 cells/scaffold). Then, 40 μL of expansion or control medium was added and the gel was spontaneously formed inside the PCL scaffolds, where the cells were embedded. After 30 min, 60 μL of medium was added in the well, and after 2 h, 700 μL was finally added. 3D cell cultures were maintained in the incubator at 37°C and 5% CO_2, and medium was changed every other day.

3. Representative results

Autologous chondrocytes are one of the most attractive cell types for CTE, due to their intrinsic properties regarding cell function, since they are found in the native cartilage. Chondrocytes are characterized by a rounded morphology, the production of tissue-specific ECM components such as collagen type I and II and glycosaminoglycans (GAGs). One of the main challenges in CTE is to obtain enough cell mass to develop a tissue construct with the desirable biological and biomechanical properties. Particularly, articular chondrocytes are obtained by invasive techniques and cell number in patient biopsies is limited. Therefore, after isolation, chondrocytes need to be expanded in 2D monolayer [1]. The expansion process leads to a rapid downregulation of chondrogenic markers, such as Collagen type I (COL1) collagen type II (COL2) and Aggrecan (ACAN) [8, 9]. Moreover, the use of extensively passaged cells leads to some degree of hypertrophy, decreased biochemical content and compromised mechanical properties [1], which is not a good indication for cartilage substitute applications.

Mesenchymal stem cells (MSCs) have generated great interest as an alternative cell source to autologous chondrocytes. MSCs are pluripotent cells with a high proliferative capacity that can be differentiated, under the appropriate microenvironment, to numerous cell lineages, such as osteogenic, adipogenic and chondrogenic [35]. MSCs can be isolated from bone marrow, adipose tissue and other sources. In particular, the adipose tissue provides and abundant reservoir of mesenchymal stem cells (adipose-derived stem cells, ADSC), which can be obtained by non-invasive surgical techniques. ADSC can undergo chondrogenic commitment in the presence of TGF-β, ascorbate, and dexamethasone combined with a 3D culture environment [35].

Three-dimensional scaffold-based cell cultures are currently used in CTE to reestablish chondrogenic phenotype of dedifferentiated chondrocytes, since they mimic more closely the natural tissue environment. On the other hand, differentiation of ADSC to cartilage-like tissue has been achieved in various 3D scaffold systems such as alginate [36], agarose [37] and collagen [38]. We report here the development of new bicomponent scaffolds based on the self-assembling peptide RAD16-I, for guiding chondrogenic differentiation of both adipose-derived stem cells (ADSC) and expanded dedifferentiated human articular chondrocytes (hAChs).

3.1 Bicomponent scaffolds made out of heparin/self-assembling peptide hydrogels

In this section, we report the development of a nanofiber scaffold with growth factor binding affinity. The strategy consisted of adding heparin moieties to the RAD16-I peptide scaffold by mixing the two components, forming a stable composite hydrogel scaffold with a natural capacity to retain HBD-containing growth factors. To evaluate the functionality of this approach for CTE applications, ADSC were cultured in the new bicomponent scaffold and induced to chondrogenic differentiation using TGFβ-1, L-ascorbic acid 2-phosphate and dexamethasone as inductors in serum-free media. 3D cultures were maintained for 4 weeks in chondrogenic or control medium, and analyzed for proteoglycan production, protein expression and mechanical properties.

During ADSC culture in the peptide scaffold RAD16-I combined with increasing concentrations of heparin (RAD/Hep), constructs cultured under chondrogenic medium—unlike constructs under control medium—became highly stained with toluidine blue, indicating a significant production of proteoglycans (**Figure 2A**). This result correlated with the aggrecan (*ACAN*) gene expression, which was only detected in constructs under chondrogenic induction (**Figure 2B**). ADSC cultured within RAD/Hep composites also produced cartilage-specific ECM proteins, such as COL1, COL2 and COL10 (**Figure 2C**). Interestingly, a single band was obtained for COL1 in 2D culture, corresponding probably to a pro-collagen intermediate (approx. 220 kDa). Different bands (ranging from 130 to 180 kDa) were obtained for COL1 in 3D constructs under chondrogenic induction. Importantly, COL2 was only detected in 3D chondro-induced cultures.

Moreover, mechanical characterization was performed over 3D chondro-induced constructs. 3D constructs, presented a storage modulus (G′) in the same order of magnitude to chicken or calf articular cartilage, but the full mechanical response of the constructs was different from native cartilage as evidenced by tan(delta) (**Figure 2D**).

3.2 Bicomponent scaffolds made out of chondroitin sulfate or decorin and self-assembling peptide hydrogels

The next strategy was based on mimicking the native cartilage ECM by adding chondroitin sulfate or decorin molecules to the nanofiber scaffold, generating thus

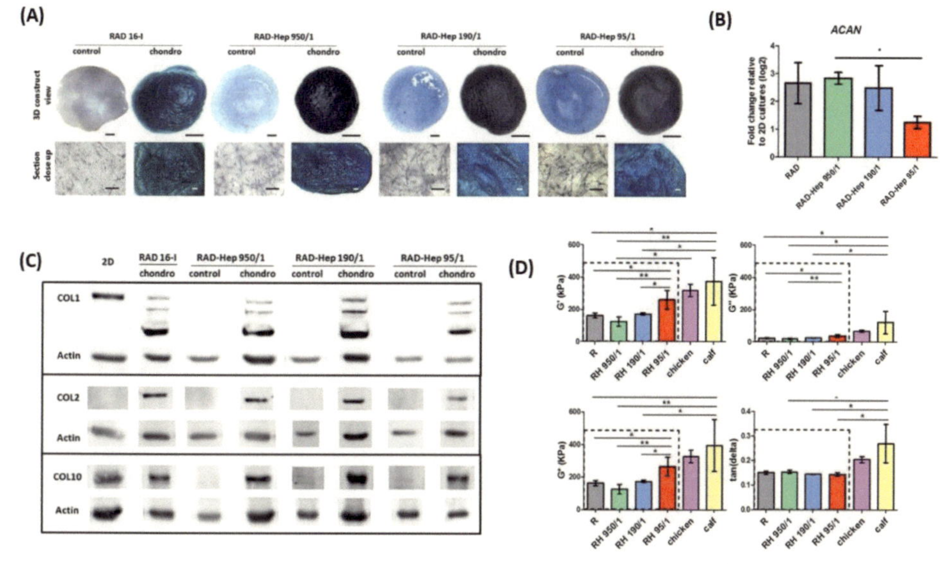

Figure 2.
Chondrogenic capacity of ADSC in RAD/heparin composite scaffold. ADSC were encapsulated within the RAD16-I peptide scaffold combined with increasing concentrations of heparin and cultured for 4 weeks under control and chondrogenic medium. (A) Toluidine blue staining of 3D ADSC constructs cultured under control and chondrogenic medium. 3D construct view scale bars = 500 μm and section close up scale bars = 100 μm. (B) Aggrecan gene expression levels of chondro-induced ADSC. Constructs cultured with control medium did not express aggrecan after 4 weeks of culture. Ct values relative to ribosomal protein L22 (RPL22) were obtained and reported as fold increase (ΔΔCt) relative to 2D cultures. (C) Protein expression characterization of ADSC cultured in RAD/Hep composites and in 2D monolayer. Western blot results of collagen type I, II and X when ADSC were maintained in control and chondrogenic medium in RAD16-I scaffold and RAD16-I/Hep composites. Actin expression was used as an internal control. (D) Mechanical characterization of 3D constructs cultured for 4 weeks in chondrogenic medium compared to chicken and calf articular cartilage. ADSCs cultured with RAD16-I and RAD/Hep composite scaffolds were analyzed for storage modulus (G', A), loss modulus (G", B), complex modulus (G, C) and tan(delta). Significant differences are indicated as *for $p < 0.05$, **for $p < 0.01$, and ***for $p < 0.001$, one-way ANOVA, N = 2 n = 3). Adapted from Fernández-Muiños et al. [25].*

chondro-favorable biochemical cues in the 3D environment. Previous work has evaluated the influence of CS to guide chondrogenesis in different hydrogel scaffolds such as chitosan [39], PEG [40], or collagen type I [41], but less is known about the ability of decorin to promote chondrogenesis commitment. In the present work, we studied the influence of both CS and decorin molecules on chondrogenesis in a nanometric 3D system. The capacity of these bicomponent scaffolds to foster chondrogenic differentiation was evaluated in two different scenarios: re-differentiation of expanded hAChs and induction of ADSC to chondrogenic commitment. Cells were seeded in RAD16-I/CS, RAD16-I/Dec and RAD16-I scaffold alone and maintained for 4 weeks in chondrogenic or control medium. Moreover, chondrocytes were also cultured in expansion medium, which contains GFs that could affect the fate of the 3D culture. 3D constructs were analyzed for morphology, gene and protein expression, proteoglycan synthesis and mechanical properties.

SEM images were obtained at week 4 of culture to assess cell morphology and their interaction with each scaffold (**Figure 3**). Articular chondrocytes cultured in expansion medium possessed a spherical morphology with possible cell-matrix interactions and thorough ECM components. Nanofibers and putative matrix components were detected on the surface of constructs cultured in control medium. Moreover, grooves with visible fibers were observed on the surface of constructs cultured in chondrogenic medium, fact that suggested the presence of secreted matrix components. On the other hand, adipose-derived stem cells under chondrogenic induction looked elongated and anchored to the scaffold surface, while

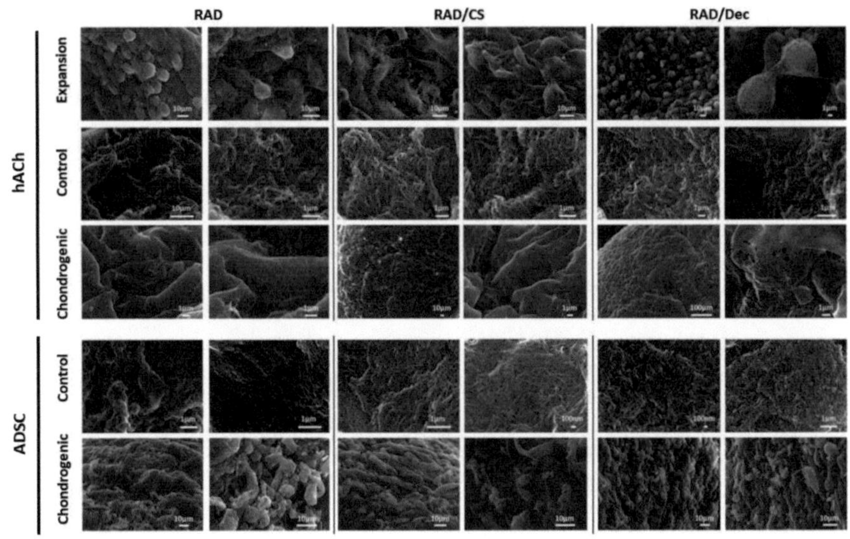

Figure 3.
SEM images of hACh and ADSC at week 4 of culture in RAD, RAD/CS and RAD/Dec scaffolds. Two images per condition are shown. Adapted from Recha-Sancho and Semino [42].

nanofibers and possible ECM components synthesized by the cells were observed in control medium (**Figure 3**). No significant differences in cell morphology were detected between RAD, RAD/CS or RAD/Dec scaffolds in any cell type.

Chondrogenic markers expression were studied at gene and protein level in hACh 3D constructs cultured in chondrogenic and expansion medium, and compared to their 2D counterparts. *COL1* was upregulated in all 3D scaffolds under chondrogenic medium and downregulated under expansion medium (**Figure 4A**). At protein level, COL1, was detected both in 2D monolayer and 3D constructs, but different band patterns were observed (**Figure 4B**). In 2D cultures, a single band was detected (approx. 220 kDa), generated probably by a pro-collagen intermediate. In 3D cultures, different bands of lower molecular weight (ranging from 130 to 180 kDa) were observed, but their intensity varied depending on the culture medium.

Interestingly, *COL2* gene expression was only upregulated in RAD/CS and RAD/Dec composite scaffolds under chondrogenic medium. This result correlated with the expression of *SOX9*, a gene regulator of *COL2*, which was significantly upregulated in 3D constructs under chondrogenic induction (**Figure 4A**). At protein level, COL2 was only detected in 3D cultures under chondrogenic induction, fact that was consistent with the gene expression profile results (**Figure 4B**). *ACAN* gene expression was higher in constructs under chondrogenic medium than in constructs cultured under expansion medium (**Figure 4A**). No differences were detected in the gene expression of hypertrophic markers compared to 2D cultures, except in RAD16-I scaffold alone, where the expression of *COL10* was upregulated in expansion medium, and *RUNX2* in chondrogenic medium (**Figure 4A**). COL10 protein expression was observed in all conditions, including 2D, but more intense bands were detected in expansion and chondrogenic medium, compared to control (**Figure 4B**).

Toluidine blue staining was performed in hACh 3D constructs to qualitatively assess the production of GAGs. Constructs under chondrogenic induction became highly stained, indicating a significant production and accumulation of GAGs by the cells (**Figure 4C**). Constructs cultured under expansion medium showed less staining, while constructs under control medium became weakly stained.

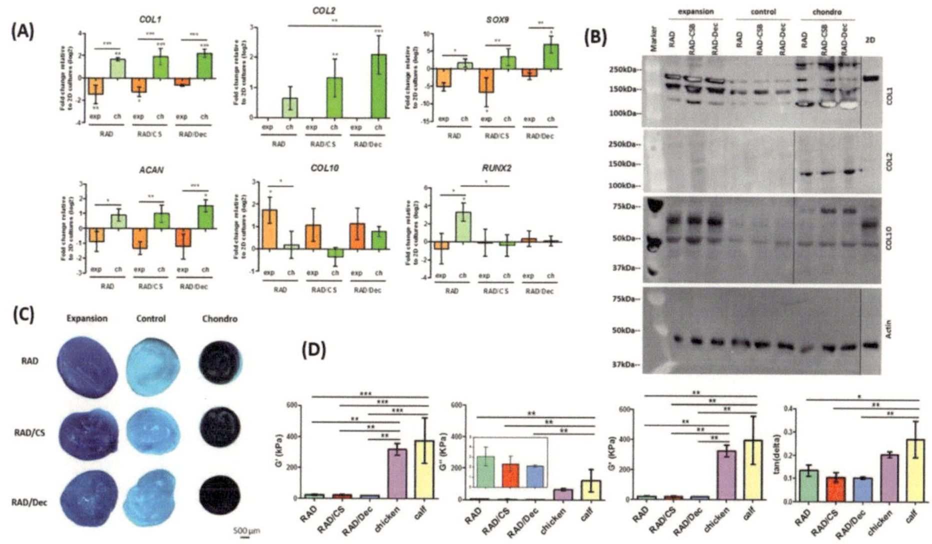

Figure 4.
Chondrogenic capacity of dedifferentiated hACh in RAD/CS and RAD/Dec 3D composite scaffolds. hACh were encapsulated within the RAD16-I peptide scaffold combined with chondroitin sulfate and decorin, and cultured for 4 weeks under expansion, control and chondrogenic medium. (A) Gene expression levels of chondrogenic and hypertrophic markers. hACh were analyzed by qRT-PCR for collagen type I (COL1), collagen type II (COL2), SOX9, aggrecan (ACAN), collagen type X (COL10) and RUNX2. Ct values relative to ribosomal protein L22 (RPL22) were obtained and reported as the fold increase ($\Delta\Delta Ct$) relative to 2D cultures (B) protein expression characterization of hACh cultured in RAD, RAD/CS and RAD/Dec composites and in 2D monolayer. Western blot results of collagen type I (COL1), II (COL2) and X (COL10) when hACh were maintained in expansion, control and chondrogenic media in the different scaffolds (RAD, RAD/CS and RAD/Dec) and in 2D monolayer. Actin expression was used as an internal control. Samples were prepared in triplicate. (C) Toluidine blue staining of hACh 3D RAD, RAD/CS and RAD/Dec constructs cultured in expansion, control and chondrogenic medium. Proteoglycan synthesis was qualitatively assessed by toluidine blue staining. (D) Mechanical characterization of 3D constructs cultured for 4 weeks in chondrogenic medium compared to chicken and calf articular cartilage. hACh cultured with RAD16-I and RAD/CS and RAD/Dec composite scaffolds were analyzed for storage modulus (G'), loss modulus (G"), complex modulus (G) and tan(delta). Significant differences are indicated as *for $p < 0.05$, **for $p < 0.01$, and ***for $p < 0.001$, one-way ANOVA, N = 2 n = 3). Adapted from Recha-Sancho and Semino [42].*

The mechanical properties of hACh 3D constructs cultured under chondrogenic medium were assessed at week 4 by dynamic mechanical analysis (DMA) and compared to calf and chicken articular cartilage (**Figure 4D**). hACh constructs exhibited lower storage modulus values (G′) than did the native cartilage samples. The viscous components (G″) and the complex modulus (G*) displayed a more similar tendency to cartilage controls. Nevertheless, all samples presented G′ values higher than G″ values, indicating that the constructs were more elastic than viscous. Tan(delta) showed that 3D constructs were comparable to chicken cartilage but differed from calf cartilage.

Chondrogenic and hypertrophic markers were studied in ADSC 3D constructs in the three scaffold types and compared to 2D monolayer culture. Results show that the gene expression of *COL1* was downregulated in 3D cultures. However, the expression of *COL2*, *SOX9* and *ACAN* was increased in 3D cultures compared to 2D. The expression levels of the hypertrophic markers *COL10* and *RUNX2* in 3D cultures were maintained at comparable levels to 2D culture (**Figure 5A**). At protein level, ADSC under chondrogenic induction produced cartilage-specific ECM proteins such as COL1, COL2 and COL10 (**Figure 5B**). As happened for hACh, one single band was obtained for *COL1* in 2D monolayer, while different bands of lower molecular weight were observed in 3D cultures. Interestingly, COL2 protein was only detected in 3D cultures.

Chondro-induced ADSC produced sulfated glycosaminoglycans, as reveals the intense staining by toluidine blue (**Figure 5C**, up). No calcium mineralization, an indicator of hypertrophy, was detected by Von Kossa staining (**Figure 5C**, down).

The mechanical properties of ADSC cultured under chondrogenic conditions in RAD, RAD/CS and RAD/Dec were assessed by dynamic mechanical analysis (DMA) at week 4 (**Figure 5D**). The constructs presented a storage modulus (G′), viscous component (G″) and complex modulus (G^*) closely related to chicken and calf cartilage. However, samples presented values of G′ much higher than G′′ so that the constructs were more elastic than viscous. Tan(delta) showed that the full mechanical response of the constructs was very similar to chicken cartilage but differed from calf cartilage.

In the present work, we aimed to induce chondrogenesis differentiation of both expanded hACh and ADSC in 3D bicomponent scaffolds made out of chondroitin sulfate or decorin and self-assembling peptide hydrogels. The expression of chondrogenic markers such as *COL2*, *SOX9* and *ACAN* was increased in both cell types compared to monolayer cultures (**Figures 4A** and **5A**). At protein level, western blot results showed a possible COL1 maturation process in 3D cultures of both cell types compared to 2D protein expression. In particular, the final mature COL1 product corresponds to the lower molecular weight band (130 kDa), which was absent in 2D cultures but predominant in constructs under chondrogenic medium (**Figures 4B** and 5B). Importantly,

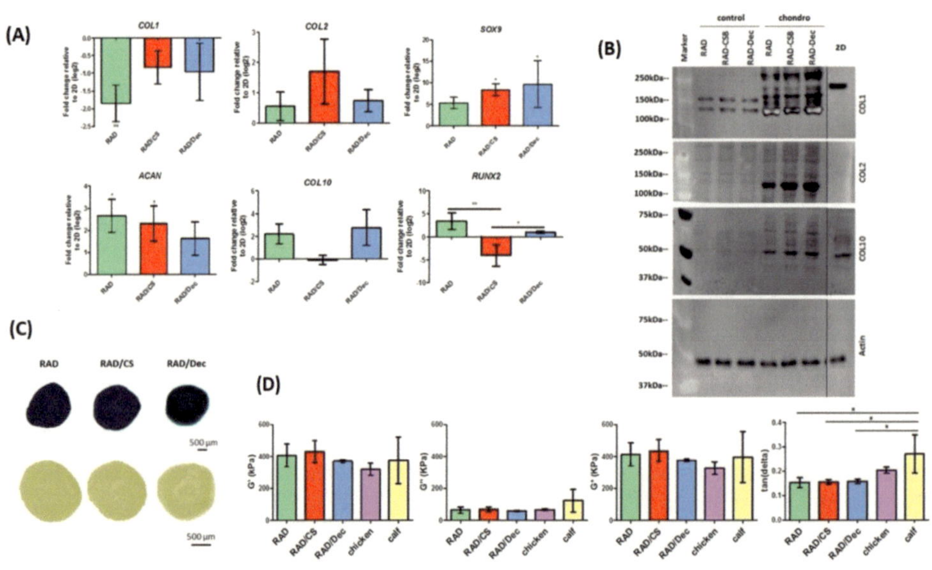

Figure 5.
Chondrogenic capacity of ADSC in RAD/CS and RAD/Dec 3D composite scaffolds. ADSC were encapsulated within the RAD, RAD/CS and RAD/Dec composite scaffolds and cultured for 4 weeks under control and chondrogenic medium. (A) Gene expression levels of chondrogenic and hypertrophic markers. ADSC were analyzed by qRT-PCR for collagen type I (COL1), collagen type II (COL2), SOX9, aggrecan (ACAN), collagen type X (COL10) and RUNX2. Ct values relative to ribosomal protein L22 (RPL22) were obtained and reported as the fold increase (ΔΔCt) relative to 2D cultures. (B) Protein expression characterization of ADSC cultured in RAD, RAD/CS and RAD/Dec composites and in 2D monolayer. Western blot results of collagen type I (COL1), II (COL2) and X (COL10) when ADSC were cultured under control and chondrogenic medium in the different scaffold types. Actin expression was used as an internal control. (C) Toluidine blue and Von Kossa staining of 3D ADSC constructs under chondrogenic induction. Proteoglycan synthesis was qualitatively assessed by toluidine blue staining (up) and calcium mineralization by Von Kossa staining (down). (D) Mechanical characterization of 3D constructs cultured for 4 weeks in chondrogenic medium compared to chicken and calf articular cartilage. ADSCs cultured with RAD16-I and RAD/CS and RAD/Dec composite scaffolds were analyzed for storage modulus (G′, A), loss modulus (G″, B), complex modulus (G, C) and tan(delta). Significant differences are indicated as * for $p < 0.05$, ** for $p < 0.01$, and *** for $p < 0.001$, one-way ANOVA, N = 2 n = 3). Adapted from Recha-Sancho and Semino [42].*

COL2 expression was only detected in 3D cultures under chondrogenic induction. Moreover, GAG production and accumulation was confirmed by toluidine blue staining (**Figures 4C** and 5C). Altogether, these results indicate the synergistic effect of the 3D culture system and the chemical inducers present in the chondrogenic medium in activating signaling pathways essentials for chondrogenic commitment, in terms of production of proteins and GAG components of the ECM. Finally, mechanical characterization showed that the viscoelastic behavior of chondro-induced ADSC constructs was more similar to native cartilage than hACh constructs (**Figures 4D** and **5D**). In resume, results until this section clearly indicate the chondro-inductive capacity of the modified scaffold which reinforce the development of biomimetic microenvironments to promote better tissue engineered cartilage substitutes.

3.3 Bicomponent scaffolds made out of PCL and self-assembling peptide hydrogels

Self-assembling peptide hydrogels provide a soft and permissive microenvironment, allowing cells to migrate, extend cellular processes and contact with other cells. Nevertheless, the use of soft hydrogels for CTE can be challenging due to its low stiffness. One approach to address this issue is the use of composite scaffolds, comprising a microscale component to increase mechanical properties and a hydrogel component (of nanoscale dimension) to promote chondrogenesis. Woven 3D poly(ε-caprolactone) (PCL) resemble native cartilage mechanical properties and, due to its high wettability, can be infiltrated with a hydrogel matrix, such as fibrin, alginate, and poly-acrylamide [43–45]. In this study, we developed a unique composite scaffold by infiltrating a 3D woven microfiber poly(ε-caprolactone) scaffold with the RAD16-I self-assembling peptide nanofiber to obtain a multi-scale functional cartilage-like tissue. The chondrogenic capacity of this new bicomponent was evaluated with expanded dedifferentiated human articular chondrocytes.

The high wettability properties of the PCL scaffold (**Figure 6A**) allowed to easily introduce the cells suspended in the RAD16-I peptide solution between the

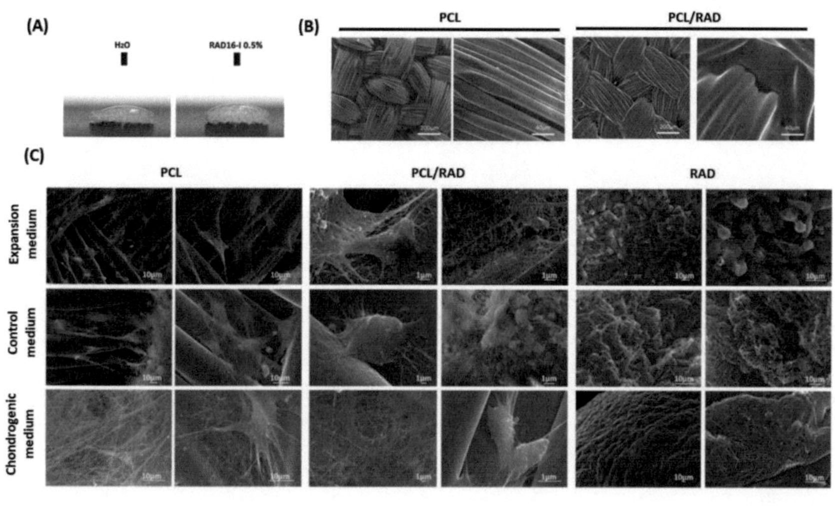

Figure 6.
SEM characterization of PCL/RAD and hACh constructs. (A) Water (left) and 0.5% RAD16-I solution (right) contact angle. The liquid was totally absorbed by the PCL scaffold (contact angle << 90°), indicating high wettability. (B) Surface view of PCL and PCL/RAD structure by SEM. 0.5% RAD16-I was lyophilized within the PCL scaffold. (C) hACh at week 4 of culture in PCL, PCL/RAD and RAD scaffolds. hACh were seeded in each scaffold and cultured in expansion, control and chondrogenic medium. Two images per condition are shown. Adapted from Recha-Sancho et al. [46].

interweaving fibers of PCL scaffold (**Figure 6B**, left). Areas of RAD16-I peptide deposition could be observed within the organized woven morphology of the fiber scaffold (**Figure 6B**, right). Thus, cells were seeded in the composite PCL/RAD and in the two scaffolds independently, PCL and RAD, and maintained for 4 weeks in expansion, chondrogenic and control medium. 3D constructs were analyzed for morphology, gene and protein expression, proteoglycan synthesis and mechanical properties.

In order to evaluate cell morphology and their interaction with the scaffolds, SEM images of hACh cultured in PCL, PCL/RAD and RAD 3D scaffolds in expansion, control and chondrogenic medium were taken at week 4 of culture (**Figure 6C**). hACh seeded in PCL scaffolds looked elongated and growing on the surface of PCL fibers. Interestingly, more fibers were detected under chondrogenic induction, probably due to an increase in extracellular matrix components production by the cells. In PCL/RAD constructs, cells seemed to be attached to the PCL fibers, with a more spherical morphology than in PCL scaffold alone, while hACh in RAD scaffolds presented in general a spherical shape.

Chondrogenic and hypertrophic markers were studied at gene and protein level at week 4 of culture in 3D scaffolds and compared to 2D cultures. *COL1* was downregulated or maintained at 2D culture levels under expansion medium, while it increased in all 3D constructs under chondrogenic conditions (**Figure 7A**). At protein level a single band (~220 kDa) was obtained for COL1 in 2D culture, while different bands of lower molecular weight (ranging from 180 to 130 kDa) were observed in 3D cultures of PCL/RAD and RAD (in all medium tested) and PCL in chondrogenic conditions (**Figure 7B**).

The expression of *COL2* was only increased in PCL/RAD and PCL scaffolds under chondrogenic induction, however, significant differences were only detected

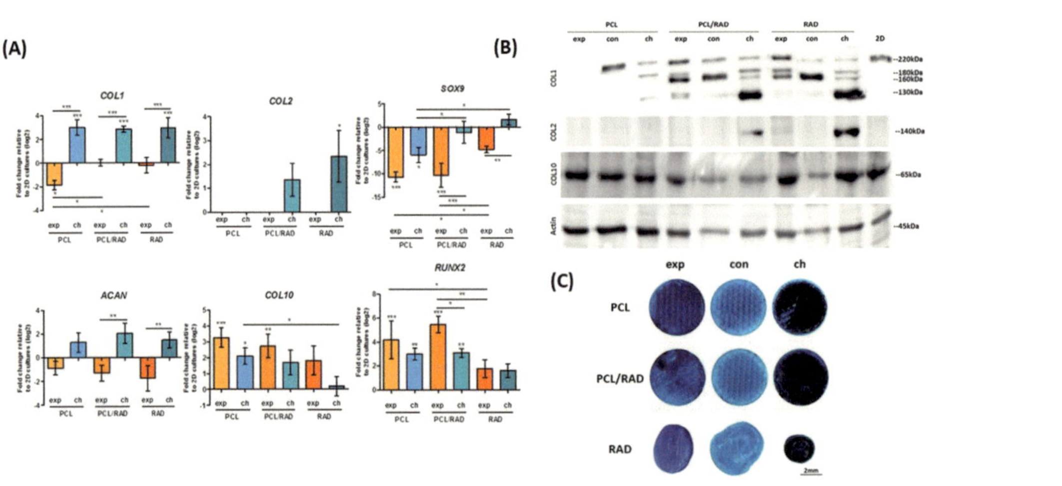

Figure 7.
*Chondrogenic capacity of dedifferentiated hACh in PCL, PCL/RAD and RAD scaffolds. hACh were seeded in each scaffold, and cultured for 4 weeks under expansion, control and chondrogenic medium. (A) Gene expression levels of chondrogenic and hypertrophic markers. hACh were analyzed by qRT-PCR for collagen type I (COL1), collagen type II (COL2), SOX9, aggrecan (ACAN), collagen type X (COL10) and RUNX2. Ct values relative to ribosomal protein L22 (RPL22) were obtained and reported as the fold increase (ΔΔCt) relative to 2D cultures. Significant differences are indicated as * for $p < 0.05$, ** for $p < 0.01$, and *** for $p < 0.001$, one-way ANOVA, N = 2 n = 3). (B) Protein expression characterization of hACh cultured in PCL, PCL/RAD and RAD scaffolds and in 2D monolayer. Western blot results of collagen type I (COL1), II (COL2) and X (COL10) when hACh were maintained in expansion, control and chondrogenic media in the different scaffolds and in 2D monolayer. Actin expression was used as an internal control. Samples were prepared in triplicate. (C) Toluidine blue staining of hACh 3D PCL, PCL/RAD and RAD constructs cultured in expansion, control and chondrogenic medium. Proteoglycan synthesis was qualitatively assessed by toluidine blue staining. Adapted from Recha-Sancho et al. [46].*

for RAD scaffold (**Figure 7A**). At protein level, COL2 was only detected in PCL/RAD and PCL scaffolds under chondrogenic medium (**Figure 7B**). *SOX9* was downregulated in PCL scaffolds in both culture medium and PCL/RAD in expansion medium. Nevertheless, it was maintained similar to 2D levels in PCL/RAD composites under chondrogenic induction and in RAD scaffold (**Figure 7A**). Aggrecan (*ACAN*) gene expression was downregulated in all scaffolds under expansion medium and upregulated in all scaffolds under chondrogenic medium, even though no differences were detected relative to 2D cultures (**Figure 7A**). Hypertrophic markers *COL10* and *RUNX2* were upregulated in some constructs respect to baseline. However, no significant increase for *COL10* was detected in RAD and PCL/RAD constructs under chondrogenic medium (**Figure 7A**). At protein level, COL10 was detected in all samples (**Figure 7B**).

The production of sulfated glycosaminoglycans was qualitatively assessed by toluidine blue staining. Constructs under chondrogenic medium were the most strongly stained compared to expansion and control medium (**Figure 7C**).

Mechanical properties of the scaffolds alone and hACh 3D constructs were assessed by dynamic mechanical analysis (DMA) at week 4 of culture, and compared to chicken and calf articular cartilage (**Figure 8**). The elastic component (G′, storage modulus) of scaffolds and 3D cultures was significantly lower than values of chicken and calf cartilage. Regarding the viscous component (G″, loss modulus), 3D constructs differed from calf native cartilage, while only PCL cellular scaffolds presented differences with chicken cartilage. All samples presented G′ values higher than G″ values, meaning that the material was more elastic than viscous. Because the complex modulus (G*) is the sum of both components, G* basically corresponds to the elastic component in this case and it presented the same pattern as the storage modulus (G′). Concerning tan(delta), which is the full mechanical response of the material, the scaffolds and cell constructs were closely related to both native cartilages, with exception of RAD constructs in chondrogenic medium, which presented differences with calf cartilage. Moreover, differences were observed between PCL/RAD and RAD constructs under the same medium. The combination of PCL scaffold and RAD hydrogel changed their viscoelastic nature after 4 weeks of culture with hACh, since tan(delta) values of the composite were increased compared to RAD scaffolds alone. This effect was not observed between composites PCL/RAD and PCL scaffold alone.

In the present study we report the chondrogenic capacity of dedifferentiated hACh in a composite scaffold comprising a microscale woven 3D poly (ε-caprolactone) and the peptide nanofiber scaffold RAD16-I. PCL scaffold resembles native cartilage mechanical properties while the RAD16-I hydrogel provides a soft and permissive 3D environment. The expression of chondrogenic markers such

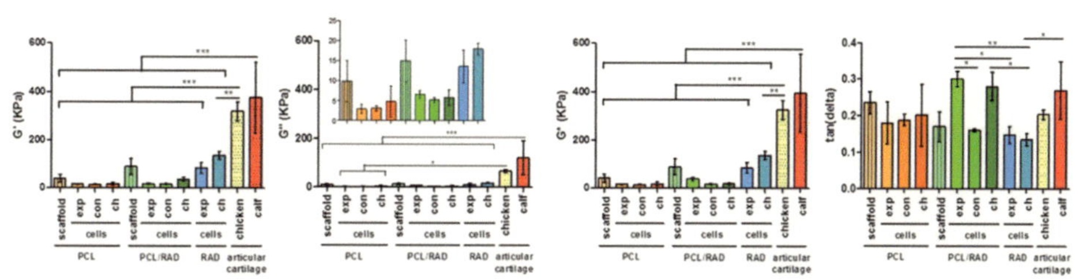

Figure 8.
Mechanical characterization of scaffolds alone and 3D constructs cultured for 4 weeks in expansion, control and chondrogenic medium compared to chicken and calf articular cartilage. hACh cultured in PCL, PCL/RAD and RAD scaffolds were analyzed for storage modulus (G'), loss modulus (G"), complex modulus (G) and tan(delta). Significant differences are indicated as *for p < 0.05, **for p < 0.01, and ***for p < 0.001, one-way ANOVA, N = 2 n = 3). Adapted from Recha-Sancho et al. [46].*

as *COL2* and *ACAN* was increased in the presence of RAD16-I peptide (in the composite and alone) compared to 2D cultures (**Figure 7A**). At protein level, different band patterns were detected for COL1, fact that suggests a protein maturation process. Specifically, the scaffolds PCL/RAD and RAD alone under chondrogenic induction, expressed higher levels of the mature COL1, as evidenced by the intensity of the 130 kDa band. Moreover, COL2 was only detected in PCL/RAD and RAD scaffolds under chondrogenic medium, suggesting that the expression of this cartilage-specific protein was due to the presence of RAD16-I hydrogel (**Figure 7B**). GAG production and accumulation was confirmed by toluidine blue staining in constructs under chondrogenic medium (**Figure 7C**). Finally, mechanical characterization showed that at the end of culture, all constructs had a viscoelastic nature (tan delta) similar to native articular cartilage, even though G′ values differed several folds from native cartilage (**Figure 8**). In resume, is clear that the combination of biomaterials to obtain a multi-dimensional composite (microfiber and nanofiber scales) is essential to acquire the best culture conditions for the cells to undergo cartilage lineage differentiation.

4. Conclusions

We report evidences from our previous work which indicates the chondro-inductive capacity of newly developed biomaterials including heparin-, chondroitin sulfate-, decorin-, and poly(ε-caprolactone)-based self-assembling peptide composite scaffolds. In particular, we demonstrated that these biomimetic biomaterials fostered re-differentiation of expanded human articular chondrocytes as well as adipose-derived stem cells into chondrogenic lineage commitment. Moreover, both biological and biomechanical properties obtained of these cartilage substitutes were comparable to natural samples of chicken and calf counterparts. This clearly suggest that these newly class of biomaterials are promising for their future application in reparative and regenerative medicine platforms.

Acknowledgements

This research was supported by funding from the AO Foundation in Acute Cartilage Injury Collaborative Research Program (ACI CRP) under the project "Bioactive and biomimetic scaffolds for cartilage regeneration" (BIOCART) and OSTEOCHON3D, and grants from the NIH (AR50245, AR48852, AG15768, AR48182, AG46927).

Conflict of interest

The authors declare no conflict of interest.

Abbreviations

2D	two dimensional
3D	three dimensional
ADSCs	adipose-derived stem cells
BMSCs	bone marrow mesenchymal stem cells
CS	chondroitin sulfate

CTE	cartilage tissue engineering
Dec	decorin
DMA	dynamic mechanical analysis
ECM	extracellular matrix
GAGs	glycosaminoglycans
G′	storage modulus
G″	loss modulus
G*	complex modulus
hACh	human articular chondrocytes
HBD	heparin binding domain
Hep	heparin
MSCs	mesenchymal stem cells
PCL	poly(ε-caprolactone)
PGs	proteoglycans
PGA	polyglycolic acid
PLGA	poly(lactic-co-glycolic acid)
PLA	polylactic acid
SAPS	self-assembling peptide scaffolds
SEM	scanning electron microscope

Author details

Nausika Betriu[2] and Carlos E. Semino[1,2]*

1 Tissue Engineering Research Laboratory, Department of Bioengineering, IQS-School of Engineering, Ramon Llull University, Barcelona, Spain

2 Hebe Biolab S.L. C/Can Castellví 27, Barcelona, Spain

*Address all correspondence to: carlos.semino@iqs.url.edu

References

[1] Chung C, Burdick JA. Engineering cartilage tissue. Advanced Drug Delivery Reviews. 2008;**60**(2):243-262. DOI: 10.1016/j.addr.2007.08.027

[2] Hunziker EB. Articular cartilage repair: Basic science and clinical progress. A review of the current status and prospects. Osteoarthritis and Cartilage. 2002;**10**(6):432-463. DOI: 10.1053/joca.2002.0801

[3] Chiang H, Jiang CC. Repair of articular cartilage dafects: Review and perspectives. Journal of the Formosan Medical Association. 2009;**108**(2):87-101. DOI: 10.1016/S0929-6646(09)60039-5

[4] Castells-sala C, Recha L. Current applications of tissue engineering in biomedicine, Journal of Biochips & Tissue Chips. 2015;**s2**:004. DOI: 10.4172/2153-0777.S2-004

[5] Cucchiarini M, Madry H, Guilak F, Saris DB, Stoddart MJ, Koon Wong M, et al. A vision on the future of articular cartilage repair. European Cells & Materials. 2014;**27**(Suppl):12-16. DOI: 10.22203/eCM

[6] Hubka KM, Dahlin RL, Meretoja VV, Kasper FK, Mikos AG. Enhancing chondrogenic phenotype for cartilage tissue engineering: Monoculture and coculture of articular chondrocytes and mesenchymal stem cells. Tissue Engineering. Part B, Reviews. 2014;**20**(6):641-654. DOI: 10.1089/ten.teb.2014.0034

[7] Pittenger MF, Mackay AM, Beck SC, Jaiswal RK, Douglas R, Mosca JD, et al. Multilineage potential of adult human mesenchymal stem cells. Science. 1999;**284**(5411):143-147. DOI: 10.1126/science.284.5411.143

[8] Darling EM, Athanasiou KA. Rapid phenotypic changes in passaged articular chondrocyte subpopulations. Journal of Orthopaedic Research. 2006;**23**(2):425-432. DOI: 10.1016/j.orthres.2004.08.008

[9] Hong E, Reddi AH. Dedifferentiation and redifferentiation of articular chondrocytes from surface and middle zones: Changes in MICRORNAS-221/-222, -140, and -143/145 expression. Tissue Engineering. Part A. 2012;**19**(7-8):1015-1022. DOI: 10.1089/ten.tea.2012.0055

[10] Griffith LG, Swartz MA. Capturing complex 3D tissue physiology in vitro. Nature Reviews. Molecular Cell Biology. 2006;**7**(3):211-224. DOI: 10.1038/nrm1858

[11] Lee S-H, Shin H. Matrices and scaffolds for delivery of bioactive molecules in bone and cartilage tissue engineering. Advanced Drug Delivery Reviews. 2007;**59**(4):339-359. DOI: 10.1016/j.addr.2007.03.016

[12] Fischbach C, Chen R, Matsumoto T, Schmelzle T, Brugge JS, Polverini PJ, et al. Engineering tumors with 3D scaffolds. Nature Methods. 2007;**4**(10):855-860. DOI: 10.1038/nmeth1085

[13] Horning JL, Sahoo SK, Vijayaraghavalu S, Dimitrijevic S, Vasir JK, Jain TK, et al. 3-D tumor model for in vitro evaluation of anticancer drugs. Molecular Pharmaceutics. 2008;**5**(5):849-862. DOI: 10.1021/mp800047v

[14] Sahoo SK, Panda AK, Labhasetwar V. Characterization of porous PLGA/PLA microparticles as a scaffold for three dimensional growth of breast cancer cells. Biomacromolecules. 2005;**6**(2):1132-1139. DOI: 10.1021/bm0492632

[15] Schneider JP, Pochan DJ, Ozbas B, Rajagopal K, Pakstis L, Kretsinger J.

Responsive hydrogels from the intramolecular folding and self-assembly of a designed peptide. Journal of the American Chemical Society. 2002;**124**(50):15030-15037. DOI: 10.1021/ja027993g

[16] Genové E, Shen C, Zhang S, Semino CE. The effect of functionalized self-assembling peptide scaffolds on human aortic endothelial cell function. Biomaterials. 2005;**26**(16):3341-3351. DOI: 10.1016/j.biomaterials.2004.08.012

[17] Semino CE. Self-assembling peptides: From bio-inspired materials to bone regeneration. Journal of Dental Research. 2008;**87**(7):606-616. DOI: 10.1177/154405910808700710

[18] Semino CE. Can we build artificial stem cell compartments ? 2003;**3**:164-169. DOI: 10.1155/S1110724303208019

[19] Elsa G, Stephanie S, Ana S, Salvador B, Augustinus B, Griffith LG, et al. Functionalized self-assembling peptide hydrogel enhance maintenance of hepatocyte activity in vitro. Journal of Cellular and Molecular Medicine. 2009;**13**(9b):3387-3397. DOI: 10.1111/j.1582-4934.2009.00970.x

[20] Sieminski AL, Semino CE, Gong H, Kamm RD. Primary sequence of ionic self-assembling peptide gels affects endothelial cell adhesion and capillary morphogenesis. Journal of Biomedical Materials Research Part A. 2008;**87**(2):494-504. DOI: 10.1002/jbm.a.31785

[21] Wu J, Marí-Buyé N, Muiños TF, Borrós S, Favia P, Semino CE. Nanometric self-assembling peptide layers maintain adult hepatocyte phenotype in sandwich cultures. Journal of Nanobiotechnology. 2010;**8**(1):29. DOI: 10.1186/1477-3155-8-29

[22] Dégano IR, Quintana L, Vilalta M, Horna D, Rubio N, Borrós S, et al. The effect of self-assembling peptide nanofiber scaffolds on mouse embryonic fibroblast implantation and proliferation. Biomaterials. 2009;**30**(6):1156-1165. DOI: 10.1016/j.biomaterials.2008.11.021

[23] Garreta E, Genové E, Salvador B, Semino CE. Osteogenic differentiation of mouse embryonic stem cells and mouse embryonic fibroblasts in a three-dimensional self-assembling peptide scaffold. Tissue Engineering. 2006;**12**(8):2215-2227. DOI: 10.1089/ten.2006.12.2215

[24] Castells-Sala C, Recha-Sancho L, Llucià-Valldeperas A, Soler-Botija C, Bayes-Genis A, Semino CE. Three-dimensional cultures of human subcutaneous adipose tissue-derived progenitor cells based on RAD16-I self-assembling peptide. Tissue Engineering. Part C, Methods. 2016;**22**(2):ten.tec.2015.0270. DOI: 10.1089/ten.tec.2015.0270

[25] Fernández-Muiños T, Recha-Sancho L, López-Chicón P, Castells-Sala C, Mata A, Semino CE. Bimolecular based heparin and self-assembling hydrogel for tissue engineering applications. Acta Biomaterialia. 2015;**16**(1):35-48. DOI: 10.1016/j.actbio.2015.01.008

[26] Chen W-C, Yao C-L, Chu I-M, Wei Y-H. Compare the effects of chondrogenesis by culture of human mesenchymal stem cells with various type of the chondroitin sulfate C. Journal of Bioscience and Bioengineering. 2011;**111**(2):226-231. DOI: 10.1016/j.jbiosc.2010.10.002

[27] Seidler DG, Dreier R. Decorin and its galactosaminoglycan chain: Extracellular regulator of cellular function? IUBMB Life. 2008;**60**(11):729-733. DOI: 10.1002/iub.115

[28] Trowbridge JM, Gallo RL. Dermatan sulfate: New functions from an old

glycosaminoglycan. Glycobiology. 2002;**12**(9):117R-125R. DOI: 10.1093/glycob/cwf066

[29] Moutos FT, Freed LE, Guilak F. A biomimetic three-dimensional woven composite scaffold for functional tissue engineering of cartilage. Nature Materials. 2007;**6**:162. DOI: 10.1038/nmat1822

[30] Huang M-H, Li S, Hutmacher DW, Coudane J, Vert M. Degradation characteristics of poly(ϵ-caprolactone)-based copolymers and blends. Journal of Applied Polymer Science. 2006;**102**(2):1681-1687. DOI: 10.1002/app.24196

[31] Rohner D, Hutmacher DW, Cheng TK, Oberholzer M, Hammer B. In vivo efficacy of bone-marrow-coated polycaprolactone scaffolds for the reconstruction of orbital defects in the pig. Journal of Biomedical Materials Research Part B: Applied Biomaterials. 2003;**66B**(2):574-580. DOI: 10.1002/jbm.b.10037

[32] Freed LE, Engelmayr GC, Borenstein JT, Moutos FT, Guilak F. Advanced material strategies for tissue engineering scaffolds. Advanced Materials. 2009;**21**(32-33):3410-3418. DOI: 10.1002/adma.200900303

[33] Baugé C, Girard N, Lhuissier E, Bazille C, Boumediene K. Regulation and role of TGFβ signaling pathway in aging and osteoarthritis joints. Aging and Disease. 2014;**5**(6):394-405. DOI: 10.14336/AD.2014.0500394

[34] Cals FLJ, Hellingman CA, Koevoet W, Baatenburg de Jong RJ, van Osch GJVM. Effects of transforming growth factor-β subtypes on in vitro cartilage production and mineralization of human bone marrow stromal-derived mesenchymal stem cells. Journal of Tissue Engineering and Regenerative Medicine. 2011;**6**(1):68-76. DOI: 10.1002/term.399

[35] Chen WH, Lai MT, Wu ATH, Wu CC, Gelovani JG, Lin CT, et al. In vitro stage-specific chondrogenesis of mesenchymal stem cells committed to chondrocytes. Arthritis and Rheumatism. 2009;**60**(2):450-459. DOI: 10.1002/art.24265

[36] Chung C, Burdick JA. Engineering cartilage tissue. Advanced Drug Delivery Reviews. 2008;**60**(2):243-262. DOI: 10.1016/j.addr.2007.08.027

[37] Awad HA, Wickham MQ, Leddy HA, Gimble JM, Guilak F. Chondrogenic differentiation of adipose-derived adult stem cells in agarose, alginate, and gelatin scaffolds. Biomaterials. 2004;**25**(16):3211-3222. DOI: 10.1016/j.biomaterials.2003.10.045

[38] Masuoka K, Asazuma T, Hattori H, et al. Tissue engineering of articular cartilage with autologous cultured adipose tissue-derived stromal cells using atelocollagen honeycomb-shaped scaffold with a membrane sealing in rabbits. Journal of Biomedical Materials Research. Part B, Applied Biomaterials. 2006;**79**(1):25-34. DOI: 10.1002/jbmb

[39] Sechriest VF, Miao YJ, Niyibizi C, Westerhausen-Larson A, Matthew HW, Evans CH, et al. GAG-augmented polysaccharide hydrogel: A novel biocompatible and biodegradable material to support chondrogenesis. Journal of Biomedical Materials Research. 2000;**49**(4):534-541. DOI: 10.1002/(SICI)1097-4636(20000315)49:4<534::AID-JBM12>3.0.CO;2-#

[40] Varghese S, Hwang NS, Canver AC, Theprungsirikul P, Lin DW, Elisseeff J. Chondroitin sulfate based niches for chondrogenic differentiation of mesenchymal stem cells. Matrix Biology. 2008;**27**(1):12-21. DOI: 10.1016/j.matbio.2007.07.002

[41] Mullen LM, Best SM, Ghose S, Wardale J, Rushton N, Cameron RE.

Bioactive IGF-1 release from collagen–GAG scaffold to enhance cartilage repair in vitro. Journal of Materials Science. Materials in Medicine. 2015;**26**(1):1-8. DOI: 10.1007/s10856-014-5325-y

[42] Recha-Sancho L, Semino CE. Chondroitin sulfate- and decorin-based self-Assembling scaffolds for cartilage tissue engineering. PLoS One. 2016;**11**(6):1-23. DOI: 10.1371/journal.pone.0157603

[43] Valonen PK, Moutos FT, Kusanagi A, Moretti MG, Diekman BO, Welter JF, et al. In vitro generation of mechanically functional cartilage grafts based on adult human stem cells and 3D-woven poly (ε-caprolactone) scaffolds. Biomaterials. 2010;**31**(8):2193-2200. DOI: 10.1016/j.biomaterials.2009.11.092

[44] Moutos FT, Guilak F. Functional properties of cell-seeded three-dimensionally woven poly (ε-caprolactone) scaffolds for cartilage tissue engineering. Tissue Engineering. Part A. 2009;**16**(4):1291-1301. DOI: 10.1089/ten.tea.2009.0480

[45] Liao I-C, Moutos FT, Estes BT, Zhao X, Guilak F. Composite three-dimensional woven scaffolds with interpenetrating network hydrogels to create functional synthetic articular cartilage. Advanced Functional Materials. 2013;**23**(47):5833-5839. DOI: 10.1002/adfm.201300483

[46] Recha-Sancho L, Moutos FT, Abellà J, Guilak F, Semino CE. Dedifferentiated human articular chondrocytes redifferentiate to a cartilage-like tissue phenotype in a poly(ε-caprolactone)/Self-assembling peptide composite scaffold. Materials. 2016;**9**(6). DOI: 10.3390/ma9060472

Chapter 8

Therapeutic Potential of Articular Cartilage Regeneration using Tissue Engineering Based on Multiphase Designs

Lizeth Fuentes-Mera, Alberto Camacho, Elisabeth Engel, Vanessa Pérez-Silos, Jorge Lara-Arias, Iván Marino-Martínez and Víctor Peña-Martínez

Abstract

Articular cartilage tissue possesses poor ability to regenerate; as the lesion progresses, it extends to the underlying subchondral bone and an osteochondral (OC) defect appears complicating the therapeutic approaches. Cartilage tissue engineering has become a very active research area capable of contributing to medical technology innovation. In this regard, the development of new biomaterials in combination with cells represents one of the best alternatives for the treatment of OC injuries. In the last decades, the strategies have been designed without considering the cartilage as a complex tissue with a functionally stratified three-dimensional structure. Today, efforts are focused on creating a starting point in the process of cartilage formation with the development of a multiphase implants that recapitulates the cartilage as an OC unit, which improves its integration. This chapter will focus on a review of tissue engineering based on multiphase designs for cartilage and OC injuries, highlighting the importance of the biomaterial selection, and also the relevance of a biomimetic approach to reach a suitable microenvironment for the differentiation and maturation of the chondral tissue.

Keywords: osteochondral regeneration, cartilage tissue engineering, multiphasic designs, biofunctionalization, vascularization

1. Introduction

Clinically, an osteochondral trauma injury usually occurs in the part of the load-bearing of the knee and ankle joint. In the particular case of the knee, as a result, in most animal models, osteochondral defects are created in the femoral condyles (**Figure 1**), which are subject to various types of mechanical loading, such as compression, shear, and hydrostatic pressure. It is commonly accepted that critical size osteochondral defects can induce significant degenerative changes in surrounding cartilage and bone, possibly due to mechanical destabilization that originates from the region of the defect that cannot support the load [1]. In this sense and due to the intrinsic properties of the chondral tissue, the repair of osteochondral defects

IntechOpen

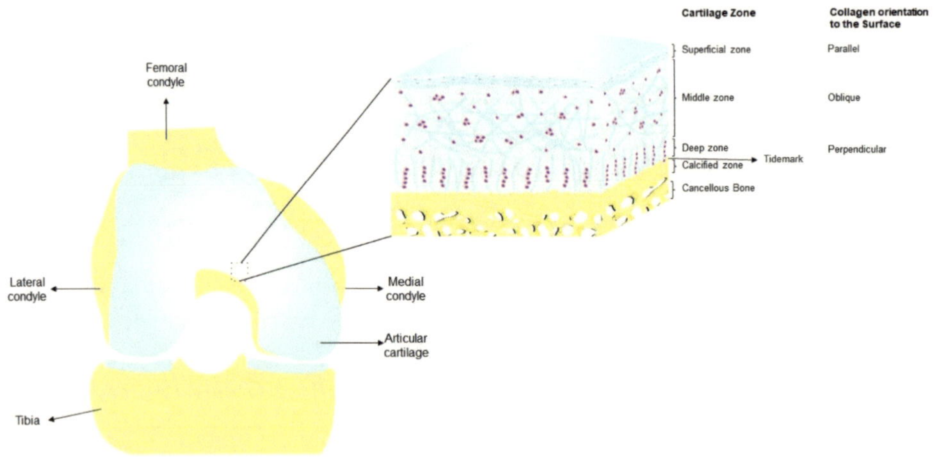

Figure 1.
Hierarchical architecture of the osteochondral unit. The layers including superficial cartilage, middle calcified cartilage, and deep subchondral bone, as well as cancellous bone, are showing; also the orientation of the collagen fibers that give cartilage its resistance compression forces.

requires an approach based on tissue engineering, so that the resulting tissue can mimic the physiological and structural properties of two different tissues (cartilage and bone) by designing specific scaffold-cell constructs. Multiphase approaches use two or three architectures, materials, and even cell types to produce a multilayer construction.

The multiphasic scaffolds have been designed to influence not only the recapitulation of the osteochondral architecture but also to make the integration of the implant with the surrounding tissue more efficient.

In the design of this type of multiphase implants, the selection of bioactive biopolymers and ceramics, but also the manufacturing method, and the dependence or not of the cellularization of the phases in harmony with the presence of the signaling factors will define the therapeutic success. This chapter aims to present and discuss the approaches currently proposed for the use of multiphase designs in the treatment of chondral and osteochondral lesions.

2. The osteochondral unit

Cartilage is a type of connective tissue whose function is to protect the bones of the diarthrodial joints from the frictional forces associated with the load and impact support [2]. Articular cartilage is predominantly avascular, aneural and alymphatic, so the main route of nutrition is through the synovial fluid and assisted by mechanical compression forces [3]. It has a variable thickness according to its location in the body; in humans, it varies from 1 to 3 mm depending on the joint. This tissue is capable of being deformed to increase the total contact surface with the consequent reduction in tension and increase the resistance to damage caused by the applied loads. This function depends on the organization of the macromolecules in the extracellular matrix, particularly the arrangement and orientation of the collagen fibers [4].

The cartilage has a single type of specialized cells called chondrocytes [5], which are embedded and grouped in the extracellular matrix (ECM) secreted by themselves. The ECM is a dynamic network of self-assembled macromolecules composed of water, gases, metabolites, cations and collagen predominantly, noncollagenous glycoproteins, hyaluronate and proteoglycans are also present. The ECM is able to regulate the behavior of cells and influences their processes of proliferation and maturation [6].

As part of the ECM, water has the function of allowing the deformation of the cartilage in response to stress; it is also important for the nourishment of the cartilage and the lubrication of the joints. Moreover, the capability of the articular cartilage to tolerate significant loads depends on the frictional resistance to water flow and the pressurization of water within the matrix. When the amount of water increases to 90%, as in osteoarthritis (OA), it causes greater permeability, which in turn causes a decrease in resistance and compromises elastic abilities [6].

The most abundant macromolecule in the ECM is collagen and represents 60% of the dry weight of the cartilage. The types of collagen present in the cartilage are I, II, IV, V, VI, IX, and XI; however, type II collagen represents 90–95% of the total amount. Collagen X, on the other hand, is only present in osteochondral ossification phases and, therefore, is associated with cartilage calcification [7].

Proteoglycans (PGs) represent 10–15% of the ECM and are the main noncollagen proteins present in the cartilage. These macromolecules are responsible for the compression of cartilage. PGs are composed of one or more linear glycosaminoglycan chains (GAGs) covalently linked [8].

At this point, we have reviewed the cellular and molecular components of joint tissue, but how are they connected to each other? The articular cartilage has a complex microarchitecture that varies from the articular surface to the subchondral bone, organized into the osteochondral unit (**Figure 1**).

The structure of the osteochondral unit is divided into four well-defined zones designated according to their morphological characteristics, that is, the content of proteoglycans or water and the density of chondrocytes in: superficial, the middle, the deep and the calcified zones (**Figure 1**). In particular, if the differences in the fibrous structure are understood, the terms "tangential," "isotropic," and "radial" have been used frequently. In consequence, the space between these zones allows, identifying three regions: the pericellular, the territorial, and the interterritorial region.

3. Histology and mechanical properties of the osteochondral unit

Each of these zones has a particular matrix composition, and cell morphology, which translates into different cellular, mechanical, and metabolic properties. It is difficult to separate the histological from the biomechanical when the cartilage is analyzed. The particular properties of loading and lubrication of articular cartilage is due, in part, to its composition, which includes a solid phase of collagen fibrils and proteoglycans entangled with a fluid phase [9]. The high tensile stiffness of the collagen considerably increases the compressive strength of the cartilage by also providing resistance to lateral expansion and allowing pressurization of the interstitial water [10]. It is believed that fluid pressurization is an important reason why articular cartilage exhibits a very low coefficient of friction [11].

As heterogeneous material consisting of surface calcified superficial layers (10–20%), medium (40–60%), and deep (30%) and thin. Each layer has specific mechanical properties and is identified by different variations in the size and direction of the collagen fibers. The content of proteoglycans is lower in the surface area and increases with depth.

The superficial area is thin and protects the deeper layers of the shear stresses. It is mainly composed of collagen types II and IX hermetically packed and in parallel alignment with the articular surface. It contains flattened chondrocytes, which are influenced by synovial fluid. This area is responsible for the traction properties of cartilage (**Figure 1**).

Below the surface area is the middle (transition) zone, which represents a bridge between the surface and deep zones. This zone contains a low density of spherical chondrocytes, proteoglycans, and fibrils of thicker collagen and is responsible for resistance to compression forces. The middle zone of the cartilage has looser collagen fibers, which gives it the greater Young compression modulus. It is these variations in tissue morphology that account for the tensile and shear strength properties of cartilage [12] (**Figure 1**).

The deep zone provides the greatest resistance to compression forces. It is formed of larger diameter collagen fibrils in a radial arrangement and a low amount of water. The chondrocytes are organized in a columnar orientation, parallel to the collagen fibers and perpendicular to the articular line (**Figure 1**).

Lastly, the calcified layer of hypertrophic chondrocytes joins the cartilage to the bone by anchoring the collagen fibrils from the deep zone of the subchondral bone (**Figure 1**) [13, 14].

Through the correlation between histology and mechanical properties, it is clear that the collagen network and the proteoglycan matrix within the articular cartilage play an important role in the control of the tensions around the chondrocytes, and in the maintenance of the good condition of the diarthrodial joints when regulating the biosynthesis of the solid matrix.

The effect of the collagen network and the fixed loading densities of the cartilage in the mechanical environment of the chondrocytes have been investigated in a depth-dependent manner. The current model emphasizes that the orientation of the collagen and the negative fixed charge densities dependent on the depth of the articular cartilage have a great effect on the modulation of the mechanical environment in the vicinity of the chondrocytes.

Apart from the structure, the composition of the cartilage is also important to determine the biomechanical properties of the tissue (e.g., traction, compression, and shearing). As mentioned above, collagen fibrils are the main contributors to the traction properties of articular cartilage. Since the different zones have different diameters of collagen and organization, the tensile properties vary significantly between the zones.

4. Clinical strategies for the osteochondral therapeutic approach

The injuries in the articular cartilage are able to stimulate a significant musculoskeletal morbidity not only in elderly patients but also in young people.

The restoration of damage from joint injuries to date represents a great challenge for medicine, since it cannot regenerate spontaneously; moreover, over time it can also lead to the establishment of osteoarthritis (OA).

The classification of articular cartilage injury is performed by instrumented palpation of the lesion and by direct observation by arthroscopy [15, 16]. The most complete classification system is established by the International Cartilage Repair Society (ICRS) [17]. The ICRS grading system evaluates the depth of the lesion and the degree to which the subchondral bone is involved to classify the injury as follows: grade 0 corresponds to a normal joint; grade 1 is presented by superficial lesions, soft cleft, and/or superficial fissures and cracks; grade 2 for abnormal lesions that extend to <50% of the depth of the cartilage; grade 3 due to serious abnormalities in which cartilage defects extend to >50% of the depth of the cartilage, as well as to the calcified layer and up to, but not through, the subchondral bone; and grade 4 for severe abnormal where there is also development of blisters in the tissue [17].

Articular cartilage has a limited capacity for repair. Injured chondrocytes (either superficial or partial thickness lesions) from the early stages develop defects in their metabolism; therefore, they are unable to maintain a normal concentration of PGs [18].

These modifications trigger the increase in tissue hydration and therefore the fibrillar disorganization of collagen [3, 19]. These changes favor an increase in the transmission of force toward the subchondral bone. By exceeding the capacity of the subchondral bone, the impact on the damaged cartilage is even deeper.

In response to this series of events, the chondrocytes proliferate and therefore the production of matrix molecules at the area of the lesion increases, however, the new matrix is not able to restore the native surface [3].

When the lesion reaches the subchondral bone (full-thickness lesions), the entry of pluripotent medullary elements is observed [20]. These migratory mesenchymal stem cells produce type I collagen fibers to fill the full thickness defect with fibrocartilage. It should be noted that fibrocartilage is not capable of supplying the damping functions of articular cartilage [21].

Following this line of argumentation, the strategies designed for the treatment of articular cartilage lesions can classically be classified as discussed below.

Palliative as physiotherapy and systemic medications to relieve pain; reparative procedures such as debridement, washing of the knee and ankle joint, arthroscopic arthroplasty, microfracture, and bone marrow stimulation techniques; restorative such as high tibial osteotomy, unicompartmental knee arthroplasty and total knee arthroplasty; and transplantation such as osteochondral transplantation (osteochondral graft), osteochondral autologous transplantation (OATS), and transplantation of a autologous chondrocyte implantation (ACI) [22, 23].

4.1 Microfracture

Classified within the reparative procedures is the microfracture. Microfracture was introduced into the clinic after other techniques of bone marrow stimulation were used in the late 1980s and early 1990s to penetrate the subchondral bone. This technique improves the migration of MSCs from the bone marrow to the site of the cartilage defect; however, microfracture often results in the formation of fibrocartilage that is biochemically and biomechanically inferior to hyaline articular cartilage [24]. A case series study has shown that without the mechanical robustness of the hyaline tissue, the repair tissue is vulnerable to joint mechanical forces and typically deteriorates between 18 and 24 months after surgery. Such deterioration is particularly evident when treating large defects or those located in the patellofemoral joint [25].

Although the FDA and many physicians still consider microfracture to be the gold standard for cartilage repair, prospective comparative studies show that microfracture could delay cartilage degeneration only in the short term; more than 5 years after surgery, treatment failure can be expected regardless of the size of the lesion [26].

4.2 Osteochondral autologous transplantation (OATS)

Osteochondral autologous transplantation has been indicated majorly for small-to-medium size (diameter > 10 mm) focal articular cartilage or osteochondral defects of the weigh-bearing areas of the femoral condyles, patellofemoral joint and talus without an acceptable outcome after less invasive techniques [27].

In OATS, a single or multiple osteochondral grafts are harvested from either the less-weight-bearing parts of the femoral condyle or the costal-osteochondral junction. This surgical procedure has the advantage of transplanting viable hyaline cartilage and subchondral bone, which is then transplanted into the defect area to restore the integrity of the articular surface [28].

The disadvantages are basically two: the availability of the grafts and the morbidity of the donor site. The major disadvantage of this procedure is the need to harvest one or multiple grafts from an asymptomatic knee or the rib area. Osteochondral harvesting in OATS often results in considerable donor-site morbidity, showing rates of 17 and 6% for ankle and knee mosaicplasty procedures, respectively, without any significant correlation between the rate of donor-site morbidity and size of the defect, number, and size of the plugs [29]. Furthermore, there is limited evidence on the short- and long-term consequences from harvesting bone plugs of asymptomatic joints.

4.3 Implantation of autologous chondrocytes

The inconsistent results of microfracture opened the way to the development of autologous chondrocyte implantation (ACI). To perform this technique, a sample of cartilage of full thickness is collected from a region of the joint under heavy weight; this by means of a biopsy during a first arthroscopic operation, the biopsy would thus serve to provide a population of chondrocytes that would later be expanded in vitro, to generate around 12–48 million cells. During a second operation, the chondrocytes would implant in the defect of the debrided cartilage to finally be covered with a membrane. This technique has two main benefits: the use of a patient's own cells, which avoids possible complications related to immune events or viral infections when transplanting allogeneic cells or foreign materials, and unlike the autologous osteochondral implantation, the small biopsy minimizes complications in the donor zone of chondrocytes [30].

The positive clinical and functional results of the ACI have been confirmed by clinical trials [31, 32]. The series of long-term cases with 5 years of follow-up have shown that ACI is an effective and durable treatment for knee cartilage lesions greater than 4 cm^2 [33, 34].

It should be noted, however, that the ACI has three main drawbacks:

- Two operations are needed; this makes the recovery time very long (6–12 months) to guarantee the maturation of the neoformed tissue and thus achieve improved clinical scores from the beginning of the study.

- The most frequently reported adverse event after ACI, using a periosteal flap to seal the cells implanted in the cartilage defect, is flap hypertrophy [33]. Therefore, alternative approaches use artificial matrices such as porcine membranes consisting of collagen mixtures types I and III or hyaluronic acid scaffolds [34–36]. These materials eventually increase the likelihood of an immune reaction, and their use is currently considered not approved in the United States.

- Preliminary studies have also shown that very often, autologous chondrocytes are "dedifferentiated" to fibrochondrocytes in culture [37]. Although other studies show that they can be redifferentiated and express chondrocytic markers after being reintroduced in an *in vitro* 3D culture system [38], large-scale cohort studies are needed to continue investigating the cost-effectiveness of the ACI in this regard.

4.4 Scaffolding-based techniques

Taking into consideration the systems that allow the grafted chondrocytes to be embedded in a three-dimensional system (3D), the osteotomy and autologous osteochondral graft transplantation has been suggested to restore normal joint congruity and minimize joint deterioration. Often, these techniques have not resulted in long-term a clinical solution, which has prompted the development of approaches that involve regenerative medicine and tissue engineering to restore articular cartilage.

The lack of a support material or scaffold to guide the synthesis and organization of the neoformed ECM could, in part, explain the variability of the results among the populations of patients treated with ACI techniques. *Ex vivo* studies have shown that successful regeneration of cartilage depends on both the proliferation rate of chondrocytes and the differentiation capacity of stem cells within a three-dimensional scaffold designed by tissue engineering; this structure then acts not only as a vehicle or cellular support but also influences the properties especially the mechanical properties of the newly formed tissue [39].

5. Tissue engineering based on multiphase designs for cartilage regeneration

Tissue engineering can be defined as the creation or induction of the formation of a specific tissue, in a specific location, through the manipulation and selection of cells, matrices, and biological stimuli. It is an interdisciplinary field that applies principles of engineering and life sciences toward the development of biological substitutes that restore, maintain, or improve tissue function [40].

Currently, tissue engineering combines the contribution of cells, undifferentiated or not, which are placed into a scaffold where growth factors can be added to accelerate cell proliferation and differentiation so that after being transplanted to a damaged structure and reaching its regeneration.

5.1 Strategies for cartilage tissue engineering

In the cartilage tissue engineering, the constant development of new designs combining biomaterials, with different cellular sources and modifying the cell culture methodology within the scaffolding systems is driven by the need, still not satisfied, to have a gold standard for functional and long-term repair of chondral and osteochondral defects.

As a cellular source for the formation of cartilage, chondrocytes, or alternatively, mesenchymal stem cells can be used. In the case of mesenchymal stem cells, there are a series of known factors that induce their differentiation toward the chondral phenotype, among which are the use of a culture medium without serum, enriched with dexamethasone, ascorbate, TGFβ, and BMPs, being the method of three-dimensional cultivation at high density one of the most used for this purpose [41, 42].

For the implantation of cells in the cartilage defect, they should be embedded in the thickness of scaffolding. These cellularized graft needs to be maintained for some period of time in culture in order to allow the cells to secrete enough ECM to functionally replace the normal cartilage and facilitate its complete integration. A newly developed osteochondral construction with inferior mechanical properties can also contribute to mechanical imbalance near the defect region until its mechanical properties have matured [43]. Mayr et al. demonstrated that the cartilage component of the osteochondral graft had only half the rigidity of the surrounding cartilage 6 months after implantation [44]. The longer the osteochondral graft takes to mature into mechanical properties, the longer the surrounding cartilage will be exposed to an excessive load, which may contribute to degenerative processes. Therefore, it is necessary to select scaffolds that allow building structures related to the biological behavior of cells into an adequate environment. In any case, for the production of cartilage, it is important to achieve the maximum production of extracellular matrix because the mechanical behavior of the implanted artificial tissue is favored, especially when using less resistant scaffolds than the normal articular cartilage.

The application of cells in scaffolds, as tissue engineering does, makes cartilage regeneration strategies complex but allows the process to be orchestrated efficiently. The critical point in these strategies is the expansion of the cells in culture in order to generate a suitable production of ECM *in vitro* and with a supportive impact on the mechanical properties *in vivo* [45].

Another challenge to overcome regarding this strategy is to achieve a competent integration of the graft after implantation. The integration of the implanted tissue with the organ requires remodeling, degradation, and at the end, formation of new tissue. The remodeling of the implanted tissue is essential for its functionality [45].

In the last decades, the strategies have been designed without considering the cartilage as a complex tissue with a functionally stratified three-dimensional structure. Today, efforts are focused on achieving a benchmark in the cartilage formation process with the development of a multiphasic implant, not only because it recapitulates the nature of native tissue, but also it takes advantage of the healing capability of bone to promote the implant integration with the surrounding tissue and then bone healing and cartilage formation. The architecture of the scaffold and the presence of migratory cells within or immediately around the graft in the bone phase of the osteochondral tissue then play a key role in the integration and therefore tissue repair.

5.2 Multiphasic scaffolding

During the last decade, there have been many new developments in various aspects of scaffolding manufacturing. Computer-aided designs and fabrication technologies are used to fabricate custom scaffolds for irregularly shaped defects [46, 47].

The materials used for scaffolds and matrices are increasingly intelligent and more versatile, and can be modified to incorporate bioactive peptides [48]. Although scaffold fabrication technologies are advancing at a rapid pace, no engineering strategy used to date can completely recapitulate the biochemical and physical characteristics of native osteochondral tissue. Although it is generally useful to simplify the approach of *in vivo* repair from an engineering point of view, for a successful *in vivo* result, the biological complexities that take place within the joint must also be taken into account in the design.

The osteochondral tissue has a heterogeneous multilayer structure composed of uncalcified cartilage (superficial, middle, and deep zone), calcified cartilage and subchondral bone.

Essentially, a multiphasic scaffold should be biocompatible able to guide the structuring of new chondral and osseous tissue, taking into account the presence and biological functionality of the interface region between them (tidemark) to achieve the mechanical properties of articular cartilage. The widespread approach uses multicomponent systems, and the exquisite melding of natural and synthetic biomaterials where the assembly strategy is fundamental since it determines the topography and the structural arrangement in which the extracellular matrix is organized, a random or a well-ordered orientation of the fibers within the chondral phase in particular.

It can be postulated that the typical lack of orientation of the collagen fiber in the repaired cartilage also has a role in the prevention of a strong integration at the level of the cartilage. The surface area of the cartilage in the normal cartilage is horizontally aligned, parallel to the direction of the joint. However, within the repaired cartilage, this provision is often lacking; therefore, the border adjacent to the native cartilage tissue and modified by genetic engineering is susceptible to rupture. The vertical orientation of collagens near the subchondral bone has been attributed to the anchoring of cartilage tissue against large strains [49]. The lack of orientation of the collagen in a dynamic loading environment of the joint probably has a role in the *in vivo* failure of the implanted constructions and the decreased integration.

Additionally, it is well documented that the rigidity of the cartilage depends on the depth, and that the superficial layer of the cartilage deforms much more than the deeper layers [50]. In this respect, when the cartilaginous component of the osteochondral scaffolds lacks the deformation patterns that vary in depth, it is likely that the levels of compression deformation mismatched between the cartilage and the implant cause a higher shear stress in the interfacial region, causing a break. Tissue engineering cartilage grafts with variable depth compressive properties have also been proposed in the past [51], and can be incorporated into future osteochondral designs.

Multiphasic scaffolds can be designed considering two or three different phases (biphasic or three phase, respectively), each of them with an architecture and composition of particular biomaterials. Since the cartilage and the subchondral bone, part of the osteochondral unit, have different biological and mechanical requirements, the first approaches in the design of multiphase implants were based on the use of two different biomaterials in order to reach a tissue-specific scaffold design; moreover, the use of different combinations of biomaterials for each phase has been reported.

Polylactic-acid (PLA)-coated polyglycolic acid (PGA) scaffold molded by the computer-aided design and manufacturing (CAD/CAM) technology proved to be ideal scaffolds for cartilage regeneration, where the presence of PLA provides adequate rigidity for the chondral phase, which is attached to polycaprolactone/hydroxyapatite (PCL/HA), an osteoconductive material, where HA generates a favorable topographical surface and biomimetic microenvironment in terms of bone tissue. The regenerated cartilage and subchondral bone showed a well-structured transition zone between the two phases, which has resulted in a better integration with the host and with mechanical properties capable of supporting the solicitation of the chondral tissue [52].

The importance of generating in the scaffold a tissue-specific microenvironment that allows the undifferentiated migrant cells of the host to find a niche for the adequate differentiation toward the chondrocytic lineage is evident.

Yun-Jeong et al. have shown that not only the microenvironment generated by the composition of the biomaterials impact on a better imitation of the osteochondral unit, but also the scaffolding structure has an important influence, being an aligned structure the most adequate in comparison with a randomly structured scaffold [55]. A stratified design of aligned channels in a biphasic scaffold using collagen type I and biphasic calcium phosphate (BCP) to mimic the cartilage and bone tissue, respectively, was manufactured by using an exquisite unidirectional freeze-casting process. Collagen is flexible, and it has specific molecular domains able to induce and support cell bioactivity [53]. Likewise, BCP provides significant osteoconductivity, bioactivity, and mechanical features [54]. However, privileging on the composition, the generation of a biphasic scaffold with a longitudinal roughness of the inner channels that serves as a guide for the correct adhesion of the cells; it results in a topography that truly emulates the osteochondral unit and shows *in vivo* a superior regeneration of the osteochondral tissue compared to the random structure [55].

Therefore, not only the pore size and porosity should be taken into account for a correct design, but also the alignment of the channels within the scaffold influencing cell migration and the proper pattern fibers of the ECM.

Likewise, multiphase can be assembled on the basis of a single biomaterial. It is possible to modify the physical properties such as roughness, pore size, and interconnectivity particularizing according to the phase, selecting a specific type of porogen and particle size, as well as through the use of different solvents and the polymerization process.

Biphasic scaffold with a cartilage phase consisting of a silk scaffold attached to a bone phase consisting of a strontium-hardystonite-gahnite (SHG)-silk scaffold has been designed by Jiao Li et al. [56]. The preparation implies a coating process of SHG ceramic scaffolds with a single silk layer using an aqueous silk fibroin solution then attached to the mixture of silk using methanol as an alternative solvent prior to silk scaffold formation in order to induce β-sheet formation in fibroin and the structure of the interface. The conformation of this design showed not only the ability to promote the differentiation of human mesenchymal stem cells toward the chondrogenic or osteogenic lineage, but also in addition, by having a well-stratified biphasic structure, the loading behavior validated the compression properties.

By the same token, a single biomaterial can be used and can generate distinct microenvironment using different molecules to biofunctionalize in a tissue-specific manner. Certainly, no biomaterial is intrinsically capable of satisfying all the requirements for the manufacture of complex and stratified tissues, so the biofunctionalization of these is presented as an alternative procedure to adapt the properties of the biomaterials to the needs of the chondral or bone tissue.

A biphasic, but monolithic scaffolds based on alginate, a highly biocompatible natural biomaterial able to support the growth of diverse cell lineages is designed by Schütz et al. through its strategy; scaffolds are fabricated by a diffusion-controlled system that allows the directed ionotropic gelation [57]. The final structure leads to the formation of channel-like, parallel aligned pores. In order to generate a chondral environment, alginate is biofunctionalized with hyaluronic acid, while for the bone phase, hydroxyapatite is used.

This simple procedure generates two well-defined layers characterized by different microstructure and mechanical properties, which provide a suitable environment for cells to form the respective tissue. Although an interface between the chondral and bone areas of the implant is not structured, a stable connection between them is clearly demonstrated, which positively impacts the mechanical properties in the final design. According to the influence of biofunctionalization, it was demonstrated by gene expression analysis that the embedded stem cells differentiated into the chondrogenic lineage when they were cultivated in chondrogenic medium; additionally, under the stimulation of the hyaluronic acid present in this phase, the chondrocyte phenotype remained stable.

Biofunctionalization, especially for monolithic scaffolds, is a useful alternative to provide chondro- and osteoinductive properties. Aragonite is a biomaterial from coral exoskeletons, similar to human bone including its 3D structure and pore interconnections as well as its crystalline form of calcium carbonate ($CaCO_3$) [58]. That features confers improved osteoconductive ability, suitable for bone regeneration.

Interestingly, specific coral species differ in size and interconnectivity of coral pores, which expands the range of applications for different tissues. In order to induce chondrogenesis in a monolithic system of aragonite, the use of hyaluronic acid has been described by Korn et al. [59]. We have already discussed before, the relevance of channel generation aligned in parallel to guide the adhesion of the cells in the chondral phase and the subsequent structuring of the ECM. In this design, in addition to the biofunctionalization with hyaluronic acid, the mechanical modification of drilled channels is also added. The combination of the two strategies showed in a model of joint damage in goat the best results compared to aragonite alone, and in the absence of parallel channels; it means a cartilaginous repair tissue with hyaline cartilage as shown by the marked expression of proteoglycans, as well as of collagen type II and absence of collagen type I.

6. Influence of vascularization on scaffold design for osteochondral regeneration

Vascularization is the bottleneck in tissue engineering. Creating constructs in the laboratory that lack of the proper vessels network will fail after implantation as the cells will not get enough oxygen and nutrients and will die. This fact is even more significant for osteochondral regeneration. Bone is a highly vascularized tissue while cartilage is avascular. When vascular networks invade cartilage surface from the subchondral zone, it might lead to an ossification of the cartilage from the deep and intermediate zone implying a joint damage and increasing pain. The design of the optimal scaffold to control angiogenesis, promoting vessel growth from preexisting ones, on the bone side and inhibiting it on the cartilage side is relevant for osteochondral regeneration.

One strategy to improve bone formation is to use growth factors (GFs) that can activate angiogenesis within the scaffold. There are several GFs involved in angiogenesis, such as vascular endothelial growth factor (VEGF), platelet-derived GF (PDGF), bone morphogenic proteins (BMPs), fibroblasts growth factors (FGFs), and tumor growth factor beta (TGFβ) [60]. Uploading VEGF is widely used, as the VEGF activates endothelial precursor cell (EPC) migration and proliferation activating the angiogenic process, and subsequently promotes the recruitment and survival of bone forming cells improving bone regeneration. However, the presence of high levels of VEGF is one of the factors related to OA progression, inducing cartilage degeneration and pain [61].

Therefore, the scaffold design for osteochondral regeneration must fulfill different properties that are shared by the two tissues, such as cell adhesion and proliferation and a high production of ECM; but others must deal with angiogenic promotion for bone or angiogenic inhibition for cartilage. Furthermore, the already observed side effects of supraphysiological doses of bone-related GFs heterotopic bone growth, pseudoarthrosis, local inflammation, and immune response [62] must be controlled by means of delivery vehicles that will ensure the bioactivity of these molecules and the remaining in the target location over the therapeutic timeframe. This can be done by covalent attachment to the scaffold, noncovalent binding, or with the nanoparticle carriers.

Kempen et al. developed a system for the sequential release of VEGF with BMP-2. BMP-2-loaded PLGA microspheres in a poly(propylene fumarate) (PPF)-scaffold combined with a VEGF-loaded gelatin hydrogel in a rat subcutaneous model demonstrated both improved vessel and bone formation when compared to scaffolds that did not contain VEGF [63].

García-Fernández et al. used antiangiogenic polymer based on 2-acrylamido-2-methylpropane sulfonic acid (AMPS) and a methacrylic derivative of 5-amino-2-naphthalenesulfonic acid (MANSA) [64]. The use of this synthetic polymer completely inhibited angiogenesis by the interaction of the sulfonic acid groups with the bFGF and VEGF modulating their activity in the processes of endothelial cell migration and proliferation. Thus, the fabrication of a biphasic scaffold by combining two different polymers that can control angiogenesis can be an efficient approach.

An innovative approach that has been tested recently is the use of microRNAs(miRNAs) to modulate cell activity for regenerative medicine applications. miRNA is a single-stranded RNA, with a length between 21 and 25 nucleotides that can regulate gene expression, usually by destabilizing mRNAs or by suppressing translation. The roles of these miRNAs on bone diseases (such as osteoporosis, osteoarthritis, and rheumatoid arthritis) have been recently highlighted. Five freely circulating miRNAs and bone tissue miRNAs are associated with osteoporotic fractures [65, 66]. miR-26a was reported to regulate angiogenesis by targeting BMP/SMAD1 signaling in endothelial cells [67].

The use of these molecules as miRNA regulators can be done by using synthetic molecules, which either mimic or repress the function of endogenous miRNAs. The mimicking molecules will enhance the suppression of the target protein synthesis by degrading the miRNA or inhibiting the protein translation. On the other hand, miRNAs inhibitors (antagomirs) preventing the activity inside the cells will lead to a rise of mRNA and protein expression. This approach can be used to upload scaffolds with either agonist or antagonist molecules to induce or avoid vascularization.

Many scaffolds have been designed to fulfill the function of miRNA delivery, mainly hydrogels, nanofibers, and porous or spongy scaffolds. Besides, the normal desired properties such as biocompatibility, easy fabrication, easy sterilization, proper mechanical properties, and adequate porosity for vessels growth, the material must retain the miRNA complexes while facilitating their sufficient exposure to the infiltrating cells without affecting its mechanical properties [68].

7. Biomaterials for multiphasic scaffolding

Biomaterial scaffold properties are fundamental to guide and recreate the native environment. The biomaterials for osteochondral applications in first insight must be biocompatible and should be intrinsically osteoinductive, osteoconductive, chondroinductive, or chondroconductive, and not less insignificant, and must possess a degradation rate that allows the formation of new tissue.

As previously stated, an ideal scaffold for the treatment from a multiphase point of view must have a chondrogenic matrix that is flexible, resistant and with pores small enough to mimic the hyaline cartilaginous matrix and an osteogenic matrix that should be mechanically competent similar to cancellous bone and bioactive, which has larger pores that mimic the microenvironment of the subchondral bone.

Achieving an articular cartilage design capable of mimicking its anisotropic mechanical behavior, still represents one of the greatest challenges in the cartilage tissues engineering [69]. In addition, the ideal biomaterial for cartilage should allow the cartilage composition to be recreated in terms of the liquid and solid phases of the connective tissue, reproduce its zonal organization, and facilitate the integration of the neoformed tissue with the adjacent native tissue.

Functionally, we can classify biomaterials into: protein-based polymers, such as fibrin, gelatin, collagen, and silk fibroin [70–74]. Biopolymers based on carbohydrates such as alginate, chitosan, agarose and polyethylene glycol [75–78], and synthetic polymers such as polylactic acid, polyglycolic acid, polycaprolactone and polylactic-coglycolicacid (PLGA) are the most common [79–81].

7.1 Carbohydrate-based polymers

These kinds of biomaterials are comprised of cross-linked polymers that swallow a great amount of water, which empathizes with the features of cartilage ECM, thus favoring the maintenance of spherical morphology within the scaffold [76]. Furthermore, synthetic materials and growth factors can be added in order to enhance chondrogenesis.

A material with adequate characteristics for cartilage engineering is chitosan, a polycationic polysaccharide that can be degraded enzymatically by the lysozyme present in the MEC of human cartilage. Chitosan has a chemical similarity with GAGs, which gives it the ability to interact with them [82]; through various *in vitro* studies, it has been demonstrated that scaffolds based on chitosan especially in combination with other biomaterials such as collagen II [108], hyaluronic acid [83], or fibroin [84] promote chondrogenic activity and support the production of aggrecan and type II collagen, thus improving cartilage repair [108].

7.2 Protein-based polymers

Among the materials of a protein nature is gelatin, which is formed from denatured collagen and can bind to growth factors, proteins, and peptides and is also capable of promoting efficient cell adhesion. On the other hand, there is the collagen that constitutes the main structural component of the ECM, and its use as a scaffolding material allows the cells to retain their phenotypes [85].

Collagen is a naturally occurring protein found in various fibrous tissues such as bone and cartilage. Collagen-based scaffolds have been used for cartilage tissue engineering applications as biomaterials due to its biocompatibility and biodegradability. Type I collagen gels seeded with bone marrow-derived MSCs have shown the formation of cartilage and subchondral bone after implantation in a full-thickness osteochondral defects macaque model. After 24 weeks, the defect had been covered with cartilage-rich reparative tissue, suitable integration with the surrounding cartilage tissue, and restoration of trabecular subchondral bone [86].

As part of this group of biomaterials is the silkworm fibroin, which is a natural biopolymer, with properties such as biocompatibility and biodegradability that allow it to be currently used for the development of a wide variety of biomedical devices and new regeneration technologies [87]. Fibroin is the main constituent (72–81%) of silkworm cocoons *Bombyx mori* [88], is a hydrophobic glycoprotein containing a large amount of hydrogen bonds, its composition and molecular orientation allows the formation of a semicrystalline structure formed by a highly ordered phase of antiparallel β-sheets that give it strength and tenacity, separated by less ordered β-sheet spacers that in turn contribute to the flexibility and elasticity of the fiber [89].

Because of these unique intrinsic properties and their versatility, silk alone is used as a biomaterial for biotechnological processes and as well as in tissue engineering [56, 90, 109]; however, it can also be combined with other polymers; the combination of fibroin/hyaluronic acid is reported, which favors the cultivation of mesenchymal stem cells [91]. In this context, silk fibroin has interesting applications in the engineering of hard and soft tissues and has diverse characteristics among which is included the ability to support the proliferation and differentiation of various cell types, making it an attractive therapeutic candidate in cartilage regenerative medicine (**Table 1**) [56, 109].

Silk fibroin has been used in several medical applications, and it can be used as fiber [92], electrospun fibers [93], films [94], or hydrogels [95]. The versatility of fibroin as a biomaterial makes it suitable for any type of application in tissue engineering, and applications that demonstrate greater maturity and close to its final application are in the field of regeneration of bone, cartilage, and ligaments.

In this regard, a very interesting application is the reconstruction of the cruciate ligament of the knee through the elaboration of a cord of silk fibers that later are sown with mesenchymal cells of the bone marrow that differentiate to ligament tissue, offering a mechanical resistance much superior to that of other organic materials and a great biocompatibility. This application is already in commercial phase in the United States, by a company specialized in the development of biomaterials based on silk fibroin (Serica) [96].

Regarding the regeneration of cartilage, fibroin has been used for the manufacture of biphasic implants in combination with bioactive ceramics or 70S bioactive glass, which

Multiphasic design	In vitro evaluation	Preclinical evaluation	Clinical evidence	References
Bilayered scaffold using microfibrillar articular cartilage extracellular matrix (ACECM) and cellularized with rabbit chondrocytes [chondral phase], attached to ACECM and nanophase hydroxyapatite (HAp) [bone phase]	A gradual interfacial region was formed; for chondral phase, intensely stained with safranin O and toluidine blue, indicating an ECM rich in sulfated proteoglycans, while bone phase a positive alizarin red staining of the lower layers indicated the rich Ca content	Not performed	Not performed	[106]
Juvenile ovine articular chondrocytes (joACs) embedded in agarose [chondral phase], attached to hydroxyapatite (HAp) ceramics [bone phase]	Suitable compressive strength due to HAp. Chondrocytes were densely packed in a GAGs and collagen-rich ECM, showing a zonal organization reminiscent of native cartilage	Not performed	Not performed	[107]
Chondrocyte cultivated on a biphasic, type II collagen–chitosan, attached to poly(lactic-co-glycolic acid) [PLGA] bone scaffold	Efficient integration between chondral and bone phases with suitable pore size differences. In vitro chondrogenic differentiation confirmed by the expression of collagen type II, Sox-9 and a remarkable upregulation of aggrecan	Not performed	Not performed	[108]
Biphasic scaffold with a cartilage phase consisting of a silk scaffold attached to a bone phase consisting of an SHG-silk scaffold (strontium- hardystonite-gahnite), and cellularized with hMSCs from bone marrow	A well-integrated interface with a stratified compressive properties according to osteochondral tissue. The structuring and maturation of the ECM congruent with the distribution and structure of the hyaline cartilage	Not performed	Not performed	[56]
Biphasic scaffold with a cartilage phase consisting of a silk scaffold well integrated to a silk-nanocalcium phosphate as a bone phase, and cellularized with rabbit bone marrow mesenchymal stromal cells	No data are presented	In a rabbit knee critical size osteochondral model. By histological and immunohistochemical analysis, cartilage regeneration and abundant presence of glycosaminoglycan and collagen II were observed. The formation of *de novo* subchondral bone and blood vessels were also observed	Not performed	[109]
A biphasic hydrogel composed by methacrylated chondroitin sulfate (CSMA) as chondral phase, and acryloyl chloride-poly(ε- caprolactone)-poly(ethylene glycol)- poly(ε-caprolactone)-acryloyl chloride (PECDA) as bone phase, with an interface of alginate; and cellularized with chondrocyte and osteoblast, respectively	Strong interfacial bonding and improved mechanical properties; highly interconnected porous structure suitable for cellular adhesion and growth in tissue-specific way	The scaffold induced a very weak inflammatory response and in a rabbit, osteochondral defect model provided a temporary structure and an adequate microenvironment for the ingrowth of osteochondral newly formed tissue	Not performed	[110]

Multiphasic design	In vitro evaluation	Preclinical evaluation	Clinical evidence	References
Biphasic acellular scaffold with a cartilage phase consisting of hyaluronic acid attached to a bone phase consisting of aragonite	Scaffold has an interconnecting porosity, with an average of 100-μm pore size, able to support human stem cell adhesion	Using a goat model of a critical osteochondral defect, the formation of hyaline cartilage and subchondral bone regeneration in the area of the lesion is demonstrated [111]	The scaffold was used on an Outerbridge grade IV promoting the mesenchymal stem-cell migration, and after 24 months, the articular surface appeared restored by MRI (Agili-C™) [112]	[111, 112]
Acellular scaffold made from polylactide-coglycolide copolymer and a bone phase containing calcium sulfate	A porous and resorbable scaffold with an osteoinductive environment due to the added calcium sulfate. The polyglycolic acid fibers, which are arranged in an orderly manner, give the structure a mechanical reinforcement [113]	Using a goat model of a critical osteochondral defect, a good filling of osteochondral defects, suitable integration with the native cartilage, and a high percentage of hyaline-like cartilage were demonstrated [114]	Reports in the literature about TruFit™ are controversial. It has been reported a poor integration of the implant with the surrounding tissue and poor bone regeneration [119, 120]. By contrast, it it has been used on an Outerbridge grade III and IV. Magnetic resonance imaging (MRI) shows an adequate integration of bone scaffolds in studied cases for more than 5 years and a sufficient restoration of the articular cartilage (Trufit™) [115]	[113–115, 119, 120]
Acellular, tri-phasic biomimetic scaffold with a cartilage phase consisting of equine type I collagen, a tidemark layer consisting of type I collagen and magnesiumhydroxyapatite (Mg-HA), attached to the bone phase consisting of a mineralized blend of type I collagen and Mg-HA	No data are presented	By sheep and horse model, the secretion of type II collagen in the cartilage region, and a uniform presence of type I collagen in the subchondral layer were evidenced. Also, the regeneration with good quality and well-integrated tissue was demonstrated [117, 118]	The scaffold has been used on an Outerbridge grade III and IV. MRI was performed and evaluated by magnetic resonance observation of cartilage repair tissue (MOCART) score. All the scores improved significantly from the baseline (Maioregen™) [116]	[116–118]

Table 1.
Multiphasic designs for osteochondral repair.

has allowed the obtaining of scaffolds with stratified properties capable of satisfying the complex and diverse regenerative requirements of the osteochondral tissue [97].

7.3 Synthetic polymer-based biomaterials

Several biodegradable and biocompatible polymers of synthetic origin have been developed for biomedical applications. The aliphatic polyesters, that is, poly (α-hydroxy esters), represent polymers that have great potential for their application in tissue regeneration. In this group are listed: poly (lactic acid) (PLA), poly (glycolic acid) (PGA), and poly (ε-caprolactone) (PCL) [12, 13]. There are three enantiomers of PLA, L-lactide, D-lactide, and mesolactide [98]. Of these, the most used are poly (L-lactide) (PLLA) and poly (D-lactide) (PDLA) [99]. Both PLLA and PDLA have a tensile strength and elongation at break (1–8%) [100, 101], its nature of slow crystallization predisposes that these materials are typically hard and brittle. *In vivo* studies have shown that highly crystalline PLLA degrades completely in 5 years, while mostly amorphous PDLLA loses strength in less than 2 months and is degraded in 1 year [102].

The material properties, degradation rates, and tissue compatibility of PLA can be modified by copolymerization with other monomers, resulting in copolymers such as poly (lactic acid-co-caprolactone) (PLGA), poly (lactic acid-co-caprolactone) (PLCL), poly (lactic acid-co-ethylene glycol) (PLEG), and poly (lactic acid-co-glutamic acid) (PLGM); this makes them biomaterials with highly adaptable properties for broad biomedical applications (**Table 1**) [108, 110, 113–115].

The most common synthetic material used for cartilage tissue engineering has been nonwoven PGA and PLA mesh. PGA has demonstrated good chondrogenesis both *in vitro* and *in vivo* [103]. A combination of a cell-free poly (L-lactic-coglycolic acid) scaffold and in situ bone marrow stem cells has been used for focal full-thickness cartilage defects in a rabbit model, demonstrating suitable integration of the implant and hyaline-like cartilage regeneration in 24 weeks [104].

These polymers have been approved by FDA: a PGA, PLA, and also polydioxanone-based copolymer; BioSeed1, BioTissue Technologies, Freiburg, Germany has been used for hyaline cartilage regeneration in human trials. This scaffold is cellularized with autologous articular chondrocytes showing improved clinical scores in human trials [105].

8. Current clinical applications of multiphase designs

The restoration of osteochondral tissue damage should be focused on the physiological features and the structure of the tissues that make it up (cartilage and bone), considering the different microenvironments that coexist in the native tissue. Through tissue engineering, multiphase designs have been developed, such as those discussed throughout this chapter that aspire to achieve this goal. Currently, the vast majority of them have been characterized in vitro; some already have an in vivo analysis in medium and large species, which brings them closer to clinical application. Although there are few multiphase designs that are currently available for a clinical application, they open an important direction for the rigorous evaluation of the designs found on this path.

The Agili-C™ CartiHeal is a biphasic scaffold, which consists in of a cartilage phase made of hyaluronic acid and a bone layer comprised by crystalline aragonite (calcium carbonate based). After in vivo trial (goat model), this acellular scaffolds evidenced the potential to recruit cells from the host tissues, and enhanced hyaline cartilage formation and subchondral bone regeneration with a continuous maturation process without deterioration of the repair tissue after 12 months implanted in critical osteochondral defects [111]. For clinical trials, only one clinical case has been reported in a 47-year-old man with an injury Outerbridge grade IV. The lesion was treated successfully and resumed normal activity after 18 months. In a follow-up at 24 months, restoration of the articular surface was demonstrated by MRI [112]. Although the results were encouraging, the occupation of the patient (athlete) could have a positive influence on the observed result, this makes it necessary to develop clinical trials in a larger number of patients under controlled conditions in order to extrapolate the benefits to a wider segment of the population.

The TruFit™ CB is an acellular scaffold made from polylactide-coglycolide copolymer and a bone phase containing calcium sulfate. The scaffold was used at first by direct implantation for the treatment of focal articular surface defects, but it showed some controversial results [113]. Several clinical studies have described a slow chondral restoration in the area of the lesion, due to poor bone repair [119], together with the poor integration of the implant with the surrounding tissue [120]. The long-term follow-up (up to 2 years) have also been controversial; however, the constant was delayed formation of the subchondral lamina [121]. Due to these clinical data, a thorough review of TruFit™ CB's design is necessary before arriving at an effective clinical application.

Maioregen™ is a triphasic biomimetic scaffold where the cartilage phase consists of equine type I collagen, an intermediate (tidemark like) layer consisting of type I collagen and magnesium-hydroxyapatite (Mg-HA), attached to the bone phase consisting of a mineralized blend of type I collagen and Mg-HA [116]. By preclinical tests on sheep and horses, it was possible to demonstrate the safety of the implant, but also that allowed the regeneration of the type II collagen-rich tissue after 6 months; this is a cell-free design [117, 118]. Throughout several clinical trials developed in such diverse populations ranging from 28 to 60 years and with a lesion size ranging from 1.5 to 6.0 cm^2, a good filling of the lesion and integration of the graft has been observed as a constant result. The evolution of the regeneration process has demonstrated the formation of subchondral bone and maturation of the chondral tissue in a period of 6 months. The evaluation by a high-resolution magnetic resonance imaging (MRI) shows the complete repair of the tissue in a period of 2 years in 66.7% of the cases treated, even in cases where the lesion involves the subchondral bone [116].

9. Conclusions

A cartilage repair treatment using tissue engineering comprises the implantation of bioabsorbable scaffolds that at first fill a chondral or osteochondral defect, then the production of cartilage repair tissue depends on the *de novo* synthesis of cartilage matrix elements. Such scaffolds support the local migration of cells (chondrogenic or osteogenic) that basically synthesize new extracellular matrix. The aim of all cartilage replacement strategies should focus on reconstruction of hyaline cartilage with its hierarchical organization; however, most of the current strategies based on monophasic designs lead to the production of fibrocartilage, which has inferior biological and mechanical characteristics compared to hyaline cartilage.

The design of multiphasic scaffolds aims at congruence with that of hierarchical nature, and from the studies that have been carried out over the past few years, it is clear that as a consequence, it substantially improves the integration of the implant with the surrounding osteochondral tissue, and positively influences the functional regeneration of both chondral and bone tissues. A vast array of multiphasic designs has been evaluated in vitro; however, only three are currently available in the clinic; the question that arises is: how to optimize the efforts to achieve a conclusive clinical application?

The use of scaffolding in order to recapitulate as much as possible the hierarchical structure seems to be not enough. The decision to cellularize or maintain a cell-free scaffold is crucial, and the answer will depend on the 3D system in a particular way; therefore, cellularization in each of the chondral and bone phases must be taken into account for the final design. On the other hand, the inclusion of an in vitro maturing time of the cellularized implant is desirable; thus, at the time of implantation, the graft has enough mechanical characteristics to support the mechanical request in the joint.

The needed to mimic the ECM on a molecular level is another main goal that demands to be taken into account, so the bioactivation of the biomaterials with elements such as synthetic materials as the ceramics (tricalcium phosphate, hydroxyapatite, and bioactive glass), or even the same decellularized tissue matrix, turns out to be a valuable tool for cartilage design, since these materials enhance the growth of a bone-like layer to support the overlying cartilage to the existing osteochondral defect.

Experimental studies are ongoing to evaluate innovative multiphase designs regarding the interaction with cells and the environment in an *in vivo* framework. *In vivo* trials using small animal models provide innovative concepts in osteochondral tissue engineering; nonetheless, to reach the development of clinical trials in humans, it is important to follow successful experiments using animal models that have loads and joint dimensions similar to humans. Animals such as sheep, pigs, and horses have surgically created defect sizes ranging from 0.29 to 0.79 cm^2 and have average human-like defects depths of about 0.68–1 cm. The body weights of these animals are also comparable or much heavier than humans, which makes them more appropriate models to predict the results in clinical trials.

Although the challenge to incorporate the use of multiphase designs to the clinic is still great, from the results observed in the wide range of studies, it is possible to conclude that tissue engineering approaches based on multiphasic scaffolds represent a promising therapeutic treatment for the regeneration of osteochondral defects. Moreover, based on the clinical results, it seems that a three-phase approach offers the most promising results with patients.

Acknowledgements

This work was partially supported by CONACYT (grant number: 233365-SALUD/2014). Special thanks to Stefano Serati for his valuable support and inspiration for the completion of this chapter.

Conflict of interest

The authors declare that they have no conflict of interest

Author details

Lizeth Fuentes-Mera[1*], Alberto Camacho[1,2], Elisabeth Engel[3,4,5], Vanessa Pérez-Silos[1], Jorge Lara-Arias[6], Iván Marino-Martínez[7] and Víctor Peña-Martínez[6]

1 Department of Biochemistry and Molecular Medicine, Faculty of Medicine, Autonomous University of Nuevo Leon (UANL), Monterrey, NL, Mexico

2 Neurometabolism Unit, Center for Research and Development in Health Sciences, Autonomous University of Nuevo Leon (UANL), Monterrey, NL, Mexico

3 Biomaterials for Regenerative Therapies Group, Institute for Bioengineering of Catalonia (IBEC), The Barcelona Institute of Science and Technology, Barcelona, Spain

4 Department of Material Science and Metallurgical Engineering, Technical University of Catalonia (UPC), Barcelona, Spain

5 CIBER en Bioingeniería, Biomateriales y Nanomedicina (CIBER-BBN), Barcelona, Spain

6 Orthopedics and Traumatology Service, University Hospital "Dr. José E. González", Autonomous University of Nuevo Leon (UANL), Monterrey, NL, Mexico

7 Experimental Therapies Unit, Center for Research and Development in Health Sciences, Autonomous University of Nuevo Leon (UANL), Monterrey, NL, Mexico

*Address all correspondence to: lizeth46@hotmail.com

References

[1] Schinhan M, Gruber M, Vavken P, Dorotka R, Samouh L, Chiari C, et al. Critical-size defect induces unicompartmental osteoarthritis in a stable ovine knee. Journal of Orthopaedic Research. 2012;**30**(2):214-220. DOI: 10.1002/jor.21521

[2] Buckwalter JA, Mankin HJ. Articular cartilage. Part I: Tissue design and chondrocyte-matrix interactions. The Journal of Bone and Joint Surgery. American Volume. 1997;**79-A**:600-611

[3] Mankin HJ. The response of articular cartilage to mechanical injury. The Journal of Bone & Joint Surgery. 1982;**64**(3):460-466

[4] Athanasiou KA, Rosenwasser MP, Buckwalter JA, Malinin TI, Mow VC. Interspecies comparisons of in situ intrinsic mechanical properties of distal femoral cartilage. Journal of Orthopaedic Research. 1991;**9**(3):330-340. DOI: 10.1002/jor.1100090304

[5] Ethier CR, Simmons CA. Introductory Biomechanics—From Cells to Organisms. 1st ed. Cambridge University Press; 2007. ISBN: 9780521841122

[6] Soo-Hyun SH, Turnbull J, Guimond S. Extracellular matrix and cell signalling: Thedynamic cooperation of integrin, proteoglycan and growth factor receptor. The Journal of Endocrinology. 2011;**209**(2):139-151. DOI: 10.1530/JOE-10-0377

[7] Maroudas A. Physiochemical properties of articular cartilage. In: Freeman MAR, editor. Adult Articular Cartilage. Kent, United Kingdom: Cambridge University Press; 1979. pp. 215-290

[8] Buckwalter JA, Rosenberg LA, Hunziker EB. In: Saltzman WM, Chien S, editors. Articular Cartilage and Knee Joint Function: Basic Science and Arthroscopy. Raven Press; 1990

[9] Lu XL, Mow VC. Biomechanics of articular cartilage and determination of material properties. Medicine & Science in Sports & Exercise. 2008;**40**(2):193-199. DOI: 10.1249/mss.0b013e31815cb1fc

[10] Park S, Krishnan R, Nicoll SB, Ateshian GA. Cartilage interstitial fluid load support in unconfined compression. Journal of Biomechanics. 2003;**36**(12):1785-1796

[11] Krishnan R, Kopacz M, Ateshian GA. Experimental verification of the role of interstitial fluid pressurization in cartilage lubrication. Journal of Orthopaedic Research. 2004;**22**(3):565-570

[12] Korhonen RK, Julkunen P, Wilson W, Herzog W. Importance of collagen orientation and depth-dependent fixed charge densities of cartilage on mechanical behavior of chondrocytes. Journal of Biomechanical Engineering. 2008;**130**(2):021003. DOI: 10.1115/1.2898725

[13] Buckwalter JA, Hunziker EB, Rosenberg LC, Coutts R, Adams M, Eyre D. Articular cartilage: Composition and structure. In: Woo SL-Y, Buckwalter JA, editors. Injury and Repair of the Musculoskeletal Soft Tissues. Park Ridge: American Academy of Orthopaedic Surgeons; 1991. pp. 405-425

[14] Bhosale AM, Richardson JB. Articular cartilage: Structure, injuries and review of management. British Medical Bulletin. 2008;**87**:77-95. DOI: 10.1093/bmb/ldn025

[15] Bauer M. The classification of articular cartilage lesions is performed through instrumented palpation of the lesion and by direct observation

by arthroscopy. Arthroscopy. 1988;**4**:97-102

[16] Peterson L, Minas T, Brittberg M, Nilsson A, Sjögren-Jansson E, Lindahl A. Two- to 9-year outcome after autologous chondrocyte transplantation of the knee. Clinical Orthopaedics and Related Research. 2000;**374**:212-234

[17] Brittberg M. The most complete classification system is established by the International Cartilage Repair Society (ICRS). The Journal of Bone and Joint Surgery. American Volume. 2003;**85-A**(2):58-69

[18] Sophia Fox AJ, Bedi A, Rodeo SA. The basic science of articular cartilage: Structure, composition, and function. Sports Health. 2009;**1**(6):461-468

[19] Lewis PB, LP MC 3rd, Kang RW, Cole BJ. Basic science and treatment options for articular cartilage injuries. The Journal of Orthopaedic and Sports Physical Therapy. 2006;**36**(10):717-727. DOI: 10.2519/jospt.2006.2175

[20] Goldberg VM, Caplan AI. Biologic restoration of articular surfaces. Instructional Course Lectures. 1999;**48**:623-627. Review

[21] Nehrer S, Spector M, Minas T. Histologic analysis of tissue after failed cartilage repair procedures. Clinical Orthopaedics and Related Research. 1999;**365**:149-162

[22] Gracitelli GC, Meric G, Briggs DT, Pulido PA, McCauley JC, Belloti JC, et al. Fresh osteochondral allograft transplantation for bipolar reciprocal osteochondral lesions of the knee. The American Journal of Sports Medicine;**2015, 43**(4):709-714. DOI: 10.1177/0363546514565770

[23] Farr J, Cole B, Dhawan A, Kercher J, Sherman S. Clinical cartilage restoration: Evolution and overview. Clinical Orthopaedics and Related Research. 2011;**499**(10):2696-2705. DOI: 10.1007/s11999-010-1764-z

[24] Bae DK, Yoon KH, Song SJ. Cartilage healing after microfracture in osteoarthritic knees. Arthroscopy. 2006;**22**(4):367-374. DOI: 10.1016/j.arthro.2006.01.015

[25] Kreuz PC, Steinwachs MR, Erggelet C, Krause SJ, Konrad G, Uhl M, et al. Results after microfracture of full-thickness chondral defects in different compartments in the knee. Osteoarthritis and Cartilage. 2006;**14**(11):1119-1125. DOI: 10.1016/j.joca.2006.05.003

[26] Goyal D, Keyhani S, Lee EH, Hui JH. Evidence-based status of microfracture technique: A systematic review of level I and II studies. Arthroscopy. 2013;**29**(9):1579-1588. DOI: 10.1016/j.arthro.2013.05.027

[27] Nishinaka N, Tsutsui H, Yamaguchi K, Uehara T, Nagai S, Atsumi T. Costal osteochondral autograft for reconstruction of advanced-stage osteochondritis dissecans of the capitellum. Journal of Shoulder and Elbow Surgery. 2014;**23**:1888-1897. DOI: 10.1016/j.jse.2014.06.047

[28] Moran CJ, Pascual-Garrido C, Chubinskaya S, Potter HG, Warren RF, Cole BJ, et al. Restoration of articular cartilage. The Journal of Bone and Joint Surgery. American Volume. 2014;**96**(4):336-344. DOI: 10.2106/JBJS.L.01329

[29] Andrade R, Vasta S, Pereira R, Pereira H, Papalia R, Karahan M, et al. Knee donor-site morbidity after mosaicplasty—A systematic review. Journal of Experimental Orthopaedics. 2016;**3**:31. DOI: 10.1186/s40634-016-0066-0

[30] Saris DB, Vanlauwe J, Victor J, Haspl M, Bohnsack M, Fortems Y, et al. Characterized chondrocyte

implantation results in better structural repair when treating symptomatic cartilage defects of the knee in a randomized controlled trial versus microfracture. The American Journal of Sports Medicine. 2008;**36**(2):235-246. DOI: 10.1177/0363546507311095

[31] Minas T, Von Keudell A, Bryant T, Gomoll AH. The John Insall Award: A minimum 10-year outcome study of autologous chondrocyte implantation. Clinical Orthopaedics and Related Research. 2014;**472**(1):41-51. DOI: 10.1007/s11999-013-3146-9

[32] Behery OA, Harris JD, Karnes JM, Siston RA, Flanigan DC. Factors influencing the outcome of autologous chondrocyte implantation: A systematic review. The Journal of Knee Surgery. 2013;**26**(3):203-211. DOI: 10.1055/s-0032-1329231

[33] Corpus KT, Bajaj S, Daley EL, Lee A, Kercher JS, Salata MJ, et al. Long-term evaluation of autologous chondrocyte implantation: Minimum 7-year follow-up. Cartilage. 2012;**3**(4):342-350. DOI: 10.1177/1947603512439460

[34] Gooding CR, Bartlett W, Bentley G, Skinner JA, Carrington R, Flanagan A. A prospective, randomised study comparing two techniques of autologous chondrocyte implantation for osteochondral defects in the knee: Periosteum covered versus type I/III collagen covered. The Knee. 2006;**13**(3):203-210. DOI: 10.1016/j.knee.2006.02.011

[35] Gomoll AH, Probst C, Farr J, Cole BJ, Minas T. Use of a type I/III bilayer collagen membrane decreases reoperation rates for symptomatic hypertrophy after autologous chondrocyte implantation. The American Journal of Sports Medicine. 2009;**37**(1):20S-23S. DOI: 10.1177/0363546509348477

[36] Marcacci M, Berruto M, Brocchetta D, Delcogliano A, Ghinelli D, Gobbi A, et al. Articular cartilage engineering with Hyalograft C: 3-year clinical results. Clinical Orthopaedics and Related Research. 2005;**435**:96-105. DOI: 10.1097/01.blo.0000165737.87628.5b

[37] Darling EM, Athanasiou KA. Rapid phenotypic changes in passaged articular chondrocyte subpopulations. Journal of Orthopaedic Research. 2005;**23**(2):425-432. DOI: 10.1016/j.orthres.2004.08.008

[38] Martinez I, Elvenes J, Olsen R, Bertheussen K, Johansen O. Redifferentiation of in vitro expanded adult articular chondrocytes by combining the hanging-drop cultivation method with hypoxic environment. Cell Transplantation. 2008;**17**(8):987-996

[39] Grigolo B, Lisignoli G, Piacentini A, Fiorini M, Gobbi P, Mazzotti G, et al. Evidence for redifferentiation of human chondrocytes grown on a hyaluronan-based biomaterial (HYAff 11): Molecular, immunohistochemical and ultrastructural analysis. Biomaterials. 2002;**23**(4):1187-1195. DOI: 10.1016/S0142-9612(01)00236-8

[40] Godbey WT, Atala A. In vitro systems for tissue engineering. Annals of the New York Academy of Sciences. 2002;**961**:10-26. DOI: 10.1111/j.1749-6632.2002.tb03041.x

[41] Lee S, Kim JH, Jo CH, Seong SC, Lee JC, Lee MC. Effect of serum and growth factors on chondrogenic differentiation of synovium-derived stromal cells. Tissue Engineering. Part A. 2009;**15**(11):3401-3415. DOI: 10.1089/ten.tea.2008.0466

[42] Toh WS, Lee EH, Cao T. Potential of human embryonic stem cells in cartilage tissue engineering and regenerative medicine. Stem Cell Reviews. 2011;7(3):544-559. DOI: 10.1007/s12015-010-9222-6

[43] Khoshgoftar M, Wilson W, Ito K, van Donkelaar C. The effect of tissue-engineered cartilage biomechanical and biochemical properties on its post-implantation mechanical behavior. Biomechanics and Modeling in Mechanobiology. 2013;**12**(1):43-54. DOI: 10.1007/s10237-012-0380-0

[44] Mayr HO, Klehm J, Schwan S, Hube R, Südkamp NP, Niemeyer P, et al. Microporous calcium phosphate ceramics as tissue engineering scaffolds for the repair of osteochondral defects: Biomechanical results. Acta Biomaterialia. 2013;**9**(1):4845-4855. DOI: 10.1016/j.actbio.2012.07.040

[45] Spector M. Biomaterials-based tissue engineering and regenerative medicine solutions to musculoskeletal problems. Swiss Medical Weekly. 2008;**136**:293-301

[46] Hung CT, Lima EG, Mauck RL, Takai E, LeRoux MA, Lu HH, et al. Anatomically shaped osteochondral constructs for articular cartilage repair. Journal of Biomechanics. 2003;**36**(12):1853-1864. DOI: 10.1016/S0021-9290(03)00213-6

[47] Swieszkowski W, Tuan BH, Kurzydlowski KJ, Hutmacher DW. Repair and regeneration of osteochondral defects in the articular joints. Biomolecular Engineering. 2007;**24**(5):489-495. DOI: 10.1016/j.bioeng.2007.07.014

[48] Grafahrend D, Heffels KH, Beer MV, Gasteier P, Möller M, Boehm G, et al. Degradable polyester scaffolds with controlled surface chemistry combining minimal protein adsorption with specific bioactivation. Nature Materials. 2011;**10**(1):67-73. DOI: 10.1038/nmat2904

[49] Shirazi R, Shirazi-Adl A, Hurtig M. Role of cartilage collagen fibrils networks in knee joint biomechanics under compression. Journal of Biomechanics. 2008;**41**(16):3340-3348. DOI: 10.1016/j.jbiomech.2008.09.033

[50] Schinagl RM, Gurskis D, Chen AC, Sah RL. Depth-dependent confined compression modulus of full-thickness bovine articular cartilage. Journal of Orthopaedic Research. 1997;**15**(4): 499-506. DOI: 10.1002/jor.1100150404

[51] Klein TJ, Malda J, Sah RL, Hutmacher DW. Tissue engineering of articular cartilage with biomimetic zones. Tissue Engineering Part B. 2009;**15**(2):143-157. DOI: 10.1089/ten.TEB.2008.0563

[52] Ding C, Qiao Z, Jiang W, Li H, Wei J, Zhou G, et al. Regeneration of a goat femoral head using a tissue-specific, biphasic scaffold fabricated with CAD/CAM technology. Biomaterials. 2013;**34**(28):6706-6716. DOI: 10.1016/j.biomaterials.2013.05.038

[53] Sartori M, Pagani S, Ferrari A, Costa V, Carina V, Figallo E, et al. A new bi-layered scaffold for osteochondral tissue regeneration: In vitro and in vivo preclinical investigations. Materials Science and Engineering C: Materials for Biological Applications. 2017;**70**(1):101-111. DOI: 10.1016/j.msec.2016.08.027

[54] Lobo SE, Livingston Arinzeh T. Biphasic calcium phosphate ceramics for bone regeneration and tissue engineering applications. Materials. 2010;**3**(2):815-826. DOI: 10.3390/ma3020815

[55] Seong Y-J, Kang I-G, Song E-H, Kim H-E, Jeong S-H. Calcium phosphate—Collagen scaffold with aligned pore channels for enhanced osteochondral regeneration. Advanced Healthcare Materials. 2017;**6**(24):1-11. DOI: 10.1002/adhm.201700966

[56] Li JJ, Kim K, Roohani-Esfahani SI, Guo J, Kaplan DL, Zreiqat H. A biphasic

scaffold based on silk and bioactive ceramic with stratified properties for osteochondral tissue regeneration. Journal of Materials Chemistry. B, Materials for Biology and Medicine. 2015;**3**(26):5361-5376. DOI: 10.1039/C5TB00353A

[57] Schütz K, Despang F, Lode A, Gelinsky M. Cell-laden biphasic scaffolds with anisotropic structure for the regeneration of osteochondral tissue. Journal of Tissue Engineering and Regenerative Medicine. 2016;**10**(5):404-417. DOI: 10.1002/term.1879

[58] Demers C, Hamdy CR, Corsi K, Chellat F, Tabrizian M, Yahia L. Natural coral exoskeleton as a bone graft substitute: A review. Bio-Medical Materials and Engineering. 2002;**12**(1):15-35

[59] Kon E, Filardo G, Robinson D, Eisman JA, Levy A, Zaslav K, et al. Osteochondral regeneration using a novel aragonite-hyaluronate bi-phasic scaffold in a goat model. Knee Surgery, Sports Traumatology, Arthroscopy. 2014;**22**(6):1452-1464. DOI: 10.1007/s00167-013-2467-2

[60] Lienemann PS, Lutolf MP, Ehrbar M. Biomimetic hydrogels for controlledbiomolecule delivery to augment bone regeneration. Advanced Drug Delivery Reviews. 2012;**64**(12):1078-1089. DOI: 10.1016/j.addr.2012.03.010

[61] Hamilton JL, Nagao M, Levine BR, Chen D, Olsen BR, Im H-J. Targeting VEGF and its receptors for the treatment of osteoarthritis and associated pain. Journal of Bone and Mineral Research. 2016;**31**(5):911-924. DOI: 10.1002/jbmr.2828

[62] Visser R, Rico-Llanos GA, Pulkkinen H, Becerra J. Peptides for bone tissue engineering. Journal of Controlled Release. 2016;**244**(Pt A):122-135. DOI: 10.1016/j.jconrel.2016.10.024

[63] Kempen DHR, Lu L, Heijink A, Hefferan TE, Creemers LB, Maran A, et al. Effect of local sequential VEGF andBMP-2 delivery on ectopic and orthotopic bone regeneration. Biomaterials. 2009;**30**(14):2816-2825. DOI: 10.1016/j.biomaterials.2009.01.031

[64] García-Fernández L, Halstenberg S, Unger RE, Aguilar MR, Kirkpatrick CJ, San Román J. Anti-angiogenic activity of heparin-like polysulfonated polymeric drugs in 3Dhuman cell culture. Biomaterials. 2010;**31**(31):7863-7872. DOI: 10.1016/j.biomaterials.2010.07.022

[65] Seeliger C, Karpinski K, Haug AT, Vester H, Schmitt A, Bauer JS, et al. Five freely circulating miRNAs and bone tissue miRNAs are associated with osteoporotic fractures. Journal of Bone and Mineral Research. 2014;**29**(8):1718-1728. DOI: 10.1002/jbmr.2175

[66] Kelch S, Balmayor ER, Seeliger C, Vester H, Kirschke JS, van Griensven M. miRNAs in bone tissue correlate to bone mineral density and circulating miRNAs are gender independent in osteoporotic patients. Scientific Reports. 2017;**7**(1):15861. DOI: 10.1038/s41598-017-16113-x

[67] Icli B, Wara AK, Moslehi J, Sun X, Plovie E, Cahill M, et al. MicroRNA-26a regulates pathological and physiological angiogenesis by targeting BMP/SMAD1 signaling. Circulation Research. 2013;**113**(11):1231-1241. DOI: 10.1161/CIRCRESAHA.113.301780

[68] Curtin CM, MencíaCastaño I, O'Brien FJ. Scaffold-based microRNA therapies in regenerative medicine and cancer. Advanced Healthcare Materials. 2018;**7**:1700695. DOI: 10.1002/adhm.201700695

[69] Chung C, Burdick JA. Engineering cartilage tissue. Advanced Drug Delivery Reviews. 2008;**60**(2):243-262 [Epub Oct 5, 2007. Review]

[70] Kaplonyi G, Zimmerman I, Frenyo AD, Farkas T, Nemes G. The use of fibrin adhesive in the repair of chondral and osteochondral injuries. Injury. 1988;**19**(4):267-272. DOI: 10.1016/0020-1383(88)90043-5

[71] Sams AE, Nixon AJ. Chondrocyte-laden collagen scaffolds for resurfacing extensive articular cartilage defects. Osteoarthritis and Cartilage. 1995;**3**(1):47-59

[72] Schneider U, Schmidt-Rohlfing B, Gavenis K, Maus U, Mueller-Rath R, Andereya S. A comparative study of 3 different cartilage repair techniques. Knee Surgery, Sports Traumatology, Arthroscopy. 2011;**19**(12):2145-2152. DOI: 10.1007/s00167-011-1460-x

[73] LaPorta TF, Richter A, Sgaglione NA, Grande DA. Clinical relevance of scaffolds for cartilage engineering. The Orthopedic Clinics of North America. 2012;**43**(2):245-254. DOI: 10.1016/j.ocl.2012.02.002

[74] Agrawal P, Pramanik K, Vishwanath V, Biswas A, Bissoyi A, Patra PK. Enhanced chondrogenesis of mesenchymal stem cells over silk fibroin/chitosan-chondroitin sulfate three dimensional scaffold in dynamic culture condition. Journal of Biomedical Materials Research. Part B, Applied Biomaterials. 2018;**106**(7):2576-2587. DOI: 10.1002/jbm.b.34074

[75] Mauck RL, Soltz MA, Wang CC, Wong DD, Chao PH, Valhmu WV, et al. Functional tissue engineering of articular cartilage through dynamic loading of chondrocyte-seeded agarose gels. Journal of Biomechanical Engineering. 2000;**122**(3):252-260. DOI: 10.1115/1.429656

[76] Buschmann MD, Gluzband YA, Grodzinsky AJ, Kimura JH, Hunziker EB. Chondrocytes in agarose culture synthesize a mechanically functional extracellular matrix. Journal of Orthopaedic Research. 1992;**10**(6):745-758. DOI: 10.1002/jor.1100100602

[77] Häuselmann HJ, Fernandes RJ, Mok SS, Schmid TM, Block JA, Aydelotte MB, et al. Phenotypic stability of bovine articular chondrocytes after longterm culture in alginate beads. Journal of Cell Science. 1994;**107**(Pt 1):17-27

[78] Mok SS, Masuda K, Häuselmann HJ, Aydelotte MB, Thonar EJ. Aggrecan synthesized by mature bovine chondrocytes suspended in alginate. Identification of two distinct metabolic matrix pools. The Journal of Biological Chemistry. 1944;**269**(52):33021-33027

[79] Barbeck M, Serra T, Booms P, Stojanovic S, Najman S, Engel E, et al. Analysis of the in vitro degradation and the in vivo tissue response to bi-layered 3D-printed scaffolds combining PLA and biphasic PLA/bioglass components—Guidance of the inflammatory response as basis for osteochondral regeneration. Bioactive Materials. 2017;**2**(4):208-223. DOI: 10.1016/j.bioactmat.2017.06.001

[80] Lohmann CH, Schwartz Z, Niederauer GG, Carnes DL Jr, Dean DD, Boyan BD. Pretreatment with platelet derived growth factor-BB modulates the ability of costochondral resting zone chondrocytes incorporated into PLA/PGA scaffolds to form new cartilage in vivo. Biomaterials. 2000;**21**(1):49-61

[81] Liu X, Liu S, Liu S, Cui W. Evaluation of oriented electrospun fibers for periosteal flap regeneration in biomimetic triphasic osteochondral implant. Journal of Biomedical Materials Research. Part B, Applied Biomaterials. 2014;**102**(7):1407-1414. DOI: 10.1002/jbm.b.33119

[82] Di Martino A, Sittinger M, Risbud MV. Chitosan: A versatile biopolymer for orthopaedic tissue-engineering. Biomaterials. 2005;**26**(30):5983-5990. DOI: 10.1016/j.biomaterials.2005.03.016

[83] Park H, Choi B, Hu J, Lee M. Injectable chitosan hyaluronic acid hydrogels for cartilage tissue engineering. Acta Biomaterialia. 2013;**9**(1):4779-4786. DOI: 10.1016/j.actbio.2012.08.033

[84] Bhardwaj N, Nguyen QT, Chen AC, Kaplan DL, Sah RL, Kundu SC. Potential of 3-D tissue constructs engineered from bovine chondrocytes/silk fibroin-chitosan for in vitro cartilage tissue engineering. Biomaterials. 2011;**32**(25):5773-5781. DOI: 10.1016/j.biomaterials.2011.04.061

[85] Ruoslahti E. RGD and other recognition sequences for integrins. Annual Review of Cell and Developmental Biology. 1996;**12**:697-715. DOI: 10.1146/annurev.cellbio.12.1.697

[86] Araki S, Imai S, Ishigaki H, Mimura T, Nishizawa K, Ueba H, et al. Improved quality of cartilage repair by bone marrow mesenchymal stem cells for treatment of an osteochondral defect in a cynomolgus macaque model. Acta Orthopaedica. 2015;**86**(1):119-126. DOI: [10.3109/17453674.2014.958807

[87] Sah M, Pramanik K. Regenerated silk fibroin from *B. mori* silk cocoon for tissue engineering applications. International journal of Environmental Science and Technology. 2010;**1**:404-408

[88] Kon'kov A, Pustovalova O, Adapov I. Biocompatible materials from regenerated silk for tissue engineering and medicinal therapy. Applied Biochemistry and Microbiology. 2010;**46**:739-744

[89] Wang Y, Kim HJ, Vunjak-Novakovic G, Kaplan DL. Stem cell-based tissue engineering with silk biomaterials. Biomaterials. 2006;**27**(36):6064-6082. DOI: 10.1016/j.biomaterials.2006.07.008

[90] Silva SS, Popa EG, Gomes ME, Olivera MB, Nayak S, Subia B, et al. Silk hydrogels from non-mulberry and mulberry silkworm cocoons processed with ionic liquids. Acta Biomaterialia. 2013;**9**(11):8972-8982. DOI: 10.1016/j.actbio.2013.06.044

[91] Garcia-Fuentes M, Meinel AJ, Hilbe M, Meinel L, Merkle HP. Silk fibroin/hyaluronan scaffolds for human mesenchymal stem cell culture in tissue engineering. Biomaterials. 2009;**30**(28):5068-5076. DOI: 10.1016/j.biomaterials.2009.06.008

[92] Li C, Vepari C, Jin H-J, Kim HJ, Kaplan DL. Electrospun silk-BMP-2 scaffolds for bone tissue engineering. Biomaterials. 2006;**27**(16):3115-3124. DOI: 10.1016/j.biomaterials.2006.01.022

[93] Xu S, Yan X, Zhao Y, Wang W, Yang YY. In vitro biocompatibility of electrospun silk fibroin mats with Schwann cells. Journal of Applied Polymer Science. 2011;**119**:3490

[94] Hofmann S, Foo CT, Rossetti F, Textor M, Vunjak-Novakovic G, Kaplan DL, et al. Silk fibroin as an organic polymer for controlled drug delivery. Journal of Controlled Release. 2006;**111**(1-2):219-227. DOI: 10.1016/j.jconrel.2005.12.009

[95] Yucel T, Cebe P, Kaplan DL. Vortex-induced injectable silk fibroin hydrogels. Biophysical Journal. 2009;**97**(7):2044-2050. DOI: 10.1016/j.bpj.2009.07.028

[96] Altman GH, Diaz F, Jakuba C, Calabro T, Horan RL, Chen J, et al. Silk-based biomaterials. Biomaterials. 2003;**24**(3):401-416. DOI: 10.1016/S0142-9612(02)00353-8

[97] JC M, Reardon PJ, Konwarh R, Knowles JC, Mandal BB. Mimicking

hierarchical complexity of the osteochondral interface using electrospun silk-bioactive glass composites. ACS Applied Materials & Interfaces. 2017;**9**(9):8000-8013. DOI: 10.1021/acsami.6b16590

[98] Madhavan Nampoothiri K, Nair NR, John RP. An overview of the recent developments in polylactide (PLA) research. Bioresource Technology. 2010;**101**:8493-8501

[99] Nair L, Laurencin C. Polymers as biomaterials for tissue engineering and controlled drug delivery. In: Lee K, Kaplan D, editors. Tissue Engineering I. Berlin Heidelberg: Springer; 2006. pp. 47-90

[100] Sarasua JR, Arraiza AL, Balerdi P, Maiza I. Crystallinity and mechanical properties of optically pure polylactides and their blends. Polymer Engineering and Science. 2005;**45**:745-753

[101] Perego G, Cella GD, Bastioli C. Effect of molecular weight and crystallinity on poly(lactic acid) mechanical properties. Journal of Applied Polymer Science. 1996;**59**:37-43

[102] Nair LS, Laurencin CT. Biodegradable polymers as biomaterials. Progress in Polymer Science. 2007;**32**:762-798

[103] Bryant SJ, Anseth KS. Journal of Biomedical Materials Research. 2002;**59**:63

[104] Shi J, Zhang X, Zeng X, Zhu J, Pi Y, Zhou C, et al. Orthopedics. 2012;**35**:665

[105] Ossendorf C, Kaps C, Kreuz PC, Burmester GR, Sittinger M, Erggelet C. Treatment of posttraumatic and focal osteoarthritic cartilage defects of the knee with autologous polymer-based three-dimensional chondrocyte grafts: 2-year clinical results. Arthritis Research & Therapy. 2007;**9**(2):R41

[106] Wang Y, Meng H, Yuan X, Peng J, Guo Q, Lu S, et al. Fabrication and in vitro evaluation of an articular cartilage extracellular matrix-hydroxyapatite bilayered scaffold with low permeability for interface tissue engineering. Biomedical Engineering Online. 2014;**13**:80. DOI: 10.1186/1475-925X-13-80

[107] Brown WE, Huey DJ, Hu JC, Athanasiou KA. Functional self-assembled neocartilage as part of a biphasic osteochondral construct. PLoS One. 2018;**13**(4):e0195261. DOI: 10.1371/journal.pone.0195261

[108] Su JY, Chen SH, Chen YP, Chen WC. Evaluation of magnetic nanoparticle-labeled chondrocytes cultivated on a type II collagen-chitosan/poly(lactic-co-glycolic) acid biphasic scaffold. International Journal of Molecular Sciences. 2017;**18**(1):pii: E87. DOI: 10.3390/ijms18010087

[109] Yan LP, Silva-Correia J, Oliveira MB, Vilela C, Pereira H, Sousa RA, et al. Bilayered silk/silk-nanoCaP scaffolds for osteochondral tissue engineering: In vitro and in vivo assessment of biological performance. Acta Biomaterialia. 2015;**12**:227-241. DOI: 10.1016/j.actbio.2014.10.021

[110] Liao J, Tian T, Shi S, Xie X, Ma Q, Li G, et al. The fabrication of biomimetic biphasic CAN-PAC hydrogel with a seamless interfacial layer applied in osteochondral defect repair. Bone Research. 2017;**5**:17018. DOI: 10.1038/boneres.2017.18

[111] Kon E, Filardo G, Shani J, Altschuler N, Levy A, Zaslav K, et al. Osteochondral regeneration with a novel aragonite-hyaluronate biphasic scaffold: Up to 12-month follow-up study in a goat model. Journal of Orthopaedic Surgery and Research. 2015;**10**:81. DOI: 10.1186/s13018-015-0211-y

[112] Kon E, Drobnic M, Davidson PA, Levy A, Zaslav KR, Robinson D. Chronic posttraumatic cartilage lesion of the knee

treated with an acellular osteochondral-regenerating implant: Case history with rehabilitation guidelines. Journal of Sport Rehabilitation. 2014;**23**(3):270-275. DOI: 10.1123/jsr.2013-0054

[113] Melton JTK, Wilson AJ, Chapman-Sheath P, Cossey AJ. TruFit CB bone plug: Chondral repair, scaffold design, surgical technique and early experiences. Expert Review of Medical Devices. 2010;**7**(3):333-341. DOI: 10.1586/erd.10.15

[114] Williams RJ, Gamradt SC. Articular cartilage repair using a resorbable matrix scaffold. Instructional Course Lectures. 2008;**57**:563-571

[115] Bugelli G, Ascione F, Dell'Osso G, Zampa V, Giannotti S. Biphasic bioresorbable scaffold (TruFit®) in knee osteochondral defects: 3-T MRI evaluation of osteointegration in patients with a 5-year minimum follow-up. Musculoskeletal Surgery. 2018;**102**(2):191-199. DOI: 10.1007/s12306-017-0522-8

[116] Kon E, Filardo G, Perdisa F, Di Martino A, Busacca M, Balboni F, et al. A one-step treatment for chondral and osteochondral knee defects: Clinical results of a biomimetic scaffold implantation at 2 years of follow-up. Journal of Materials Science. Materials in Medicine. 2014;**25**:2437-2444. DOI: 10.1007/s10856-014-5188-2

[117] Kon E, Delcogliano M, Filardo G, Fini M, Giavaresi G, Francioli S, et al. Orderly osteochondral regeneration in a sheep model using a novel nano-composite multilayered biomaterial. Journal of Orthopaedic Research. 2010;**28**(1):116-124

[118] Kon E, Mutini A, Arcangeli E, Delcogliano M, Filardo G, Nicoli Aldini N, et al. Novel nanostructured scaffold for osteochondral regeneration: Pilot study in horses. Journal of Tissue Engineering and Regenerative Medicine. 2010;**4**(4):300-308. DOI: 10.1002/term.243

[119] Barber FA, Dockery WD. A computed tomography scan assessment of synthetic multiphase polymer scaffolds used for osteochondral defective repair. Arthroscopy. 2011;**27**:60-64. DOI: 10.1016/j.arthro.2010.06.023

[120] Dell'Osso G, Bottai V, Bugelli G, Manisco T, Cazzella N, Celli F, et al. The biphasic bioresorbable scaffold (TruFit®) in the osteochondral knee lesions: Long-term clinical and assessment in 30 patients. Musculoskeletal Surgery. 2016;**100**:93-96. DOI: 10.1007/s12306-015-0383-y

[121] Verhaegen J, Clockaerts S, Van Osch G, Somville J, Verdonk P, Mertens P. Trufit plug for repair of osteochondral defects-Where is the evidence? Systematic review of literature. Cartilage. 2015;**6**:12-19. DOI: 10.1177/1947603514548890